Wild Strawberries

AND CREAM

Wild Strawberries

AND CREAM

Jo-Anne Clark Brown

A Cumberland House *Hearthside* Book

Cumberland House
Nashville, Tennessee

Published by Cumberland House Publishing, Inc., 431 Harding Industrial Drive, Nashville, Tennessee 37211.

Cover and interior design: Julie Pitkin
Photography: Jo-Anne Clark Brown

Library of Congress Cataloging-in-Publication Data

Brown, Jo-Anne Clark
   Wild Strawberries and cream / Jo-Anne Clark Brown
      p.   cm.
   Includes index.
   ISBN 1–58182–023–2 (pbk. : alk. paper)
   1. Cookery (Strawberries)   I. Title.
TX813.S9B76   1999
641.6'475—dc21

                  99–24748
                  CIP

Printed in the United States of America
1 2 3 4 5 6 7 — 03 02 01 00 99

Everyman knoweth wel inough where strawberies growe...

William Turner
The Names of Herbes (1548)

There were never strawberries
like the ones we had
that sultry afternoon
sitting in the step
of the open French window
facing each other
your knees held in mine
the blue plates in our lap
the strawberries glistening
in the hot sunlight.

Strawberries by Edwin Morgan
Edinburgh University Press

Wild Strawberries and Cream

TABLE OF CONTENTS

INTRODUCTION

It wouldn't seem like summer without strawberries, red, sweet, and juicy. The strawberry is the jewel of all summer's beautiful, bountiful cornucopia of berries. Its brilliant scarlet color, exquisite fragrance, and incomparable flavor make it a favorite around the world. It is the first edible berry to appear in the spring, celebrating the renewal of the earth after winter has flown and the earth warms to the strengthening rays of the sun. On a quiet walk in the woods we may chance upon a wild strawberry peeping out from under its tiny snow-white blossoms and protective green foliage. The delicate sweet juice of the fragile fraise du bois stains our lips and finger tips...its flavor is heavenly!

I had the good fortune to grow up in a part of the world where wild strawberries were abundant, growing just outside our door. One of my fondest memories of childhood is of picking strawberries with my grandmother. We would fill our berry pails in the warm sunshine and then rush home to enjoy with crème fraîche. The remaining berries would be made into strawberry jam that we would savor during the long, cold winter.

Every spring I look forward to seeing the first of summer's strawberries appear and always make the best use of their delicious flavor. The recipes in this book are the result of my long love affair with strawberries, using them in desserts, muffins, crêpes and sprinkled in salads. Their bright scarlet color always adds a distinct and elegant touch to an airy dessert such as Strawberry Chiffon Pie or Strawberry Soufflé, or to something more humble and equally tasty, such as hot, buttered toast with homemade strawberry jam. Most of the recipes are simple to make and the results are very rewarding. Your family, friends, and guests will praise you for talent and will ask for "more." Sitting quietly with your child or a friend, enjoying a bowl of strawberries, is one of the good things in life. The eye-catching beauty of the strawberry make it an especially attractive offering for any occasion.

The Nutritional Value of Strawberries

In addition to their natural beauty, strawberries are good to eat and are good for you, as they are a rich source of Vitamin C. For those who are conscious of caloric intake, many of the recipes contain yogurt rather than sour cream and light cream cheese is favored over the traditional variety. In recipes that call for a whipped cream garnish, simply use a substitute whipped topping in place of whipped cream.

A Brief History of the Strawberry

The botanical name for strawberry is fragaria, from the Latin fraga, meaning fragrant. It has a long and interesting history. The name strawberry is believed to be derived from the Old English "to straw," meaning to strew or scatter, perhaps because of the runners that root to form new plants. The wild strawberry bore two crops, one in June and a second one in the autumn. Wild strawberries have been an especially favorite delicacy throughout history and are known to have been transplanted from the wild and grown in ancient Roman gardens. Ancient writings, often accompanied by intricate botanical drawings, describe the beauty of wild strawberries. Ovid, a renowned Roman poet of the 1st century bc wrote of wild strawberries in his poem "Metamorphoses."

The frese rouge, the original French name for the strawberry, is believed to have appeared in France during the 13th century. Later, in the 16th century, wild strawberries were being cultivated in English gardens during the reign of Elizabeth I. Shakespeare must have been fond of strawberries and refers to them as "good strawberries" in his plays. The wild strawberry was prized for its ruby-red color at Elizabethan feast tables, as color was emphasized as a most important element of a meal at that time. Historically considered beneficial to one's health, physicians and apothecaries studied the strawberry for its medicinal qualities. The fruit, leaves, roots, and crowns were used to make ointments, tonics, and teas to relieve various ailments. It was also a favorite of enthusiastic gardeners, botanists, and horticulturists. Strawberries became a very popular and sought after fruit, and in the 1700's, gardeners began growing them commercially and selling them on the city streets in France and England.

When explorers and then settlers first came to America, they found wild strawberries abounded and soon new varieties from the colonies made their way to Europe. Strawberries from Virginia and Canada were especially prized for their large, early crops and their exceptional sweetness and fragrance. The hybridization of the strawberry is believed to have begun in the 18th century and has led to the many different varieties that are available today.

Growing Strawberries

Growing strawberries in your garden is one of life's simple, sweet pleasures. And indulging in a bowl full of fresh, juicy strawberries from your own garden, dusted with sugar and topped with thick cream, is even more delightful. Strawberries are easy to grow and require little care. They can be grown in a garden patch, in a window box, or in containers or hanging baskets on your patio or balcony. To prepare the soil for strawberry plants, simply spread two inches of composted manure over the area where you plan to grow them. Mix the manure well into the soil to a depth of about 10 inches. Soil should be capable of holding moisture in a well-drained area where water does not pool. If you plan to grow strawberries on a garden patio be certain to choose containers that have drainage holes. Choose healthy strawberry plants from your local nursery that will carry the varieties that grow well in your climate. In the garden, plant strawberry plants about twelve inches apart in single raised rows with space on each side to allow new growth or "runners" to establish roots. Allow twelve inches between rows. Plants may also be placed in a raised double row system with a furrow on each side for runners to take root. Be sure to spread the roots of the strawberry plant when planting to allow a sturdy root system to develop. Also make certain that the center or "crown" of each plant is above the soil line. Water each newly planted strawberry plant with with liquid root fertilizer available from your garden center. Once the root system is established plants should be fed with fertilizer in the spring and again in summer. Some strawberry growers recommend using tomato fertilizer to grow strawberries, but perhaps it is best to ask the local nursery to recommend the best fertilizer for the soil in your area.

Grass clippings or straw as a mulch under each strawberry plant allows the soil to remain cool and retain moisture. It also keeps weeds from robbing vital nutrients in the soil that strawberries will need for healthy growth. The mulch will also keep the berries clean and dry, preventing soil from splattering your strawberries on rainy days.

If you are planning to grow strawberries on your patio, terracotta strawberry pots are perfect as they have holes along the sides of the pot where the plants can be planted. Strawberry plants in containers should be watered frequently to keep the soil moist and allow healthy growth of the plants. Fertilizer will also be required more frequently for container plants to replace valuable nutrients in the soil that are needed for good growth.

In the first year of planting, the small white blossoms should be pinched off to allow a strong root system to become well established. Allowing good root development will help to produce abundant, healthy fruit the following spring.

There are many varieties of strawberries to choose from when planning your garden. June-bearing strawberries will produce fruit once a year in June. Be sure to allow the plant to develop a strong root system the first year by removing the strawberry blossoms. Preventing the growth of fruit in the first year will ensure a healthy crop of strawberries the next spring.

Everbearing strawberries produce an abundant crop during summer and a smaller crop in the autumn. If you choose everbearing plants, remove the white flowers until July to allow roots to become well established. Again, it is best to seek the advice of a garden center or nursery to determine which varieties of strawberries do well in your area. You may wish to plant a June-bearing variety and an everbearing plant to be certain that there will be plenty of strawberries in the garden all summer long!

The Care and Handling of Strawberries

Strawberries are fragile, highly perishable, and should be handled with care. Once they have been picked they will remain fresh for only two or three days in the refrigerator. Rinse gently just before you are ready to use them. Strawberries freeze well and may be stored in the freezer for one year. Choose only firm, fresh berries for freezing. Do not rinse strawberries before freezing as this can be done when you are ready to use them. To preserve nutrients, texture, and freshness of the berries, simply place them on a cookie sheet, cover with plastic wrap, and place in the freezer until firmly frozen. The frozen berries may then be placed in plastic freezer bags or containers and returned to the freezer. On a cold, midwinter day you may yearn for the taste of strawberries from your garden and your wish will be granted in a moment.

Wild Strawberries

AND CREAM

Breakfasts
and
Brunches

Strawberry Summer Soup

VERY COLORFUL AND DELICIOUS!

2 cups (500 mL) strawberries, hulled
1 cup (250 mL) raspberries
1/2 cup (125 mL) blueberries
1/2 cup (125 mL) cherries, pitted
1/2 cup (125 mL) fresh orange juice
1/2 teaspoon (2 mL) cinnamon
Sugar to taste

Wash the fruit thoroughly under cold water. Cut the strawberries in half, and the large ones in quarters. Reserve a few berries for garnish. In a blender place all berries with the orange juice and cinnamon. Purée the fruit mixture until completely blended. To remove the seeds, pour the puréed fruit through a sieve into a large glass bowl. Refrigerate until thoroughly chilled.

When ready to serve divide the chilled soup into attractive soup bowls. Garnish with reserved berries. Serve with plain yogurt if desired.

Serves 4 to 6.

Strawberry Cream Soup

SERVE CHILLED...DELICIOUS!

4 cups (1 L) fresh strawberries
2 cups (500 mL) cold water
Sugar to taste
2 egg yolks, beaten
½ cup (125 mL) light sour cream or yogurt

In a blender combine all of the ingredients. Blend a few seconds, just until well mixed. Refrigerate until chilled.

Serves 6.

Strawberry Breakfast Puff

A SPECIAL BRUNCH FOR TWO!

3 eggs
½ cup (125 mL) flour
½ cup (125 mL) buttermilk
3 tablespoons (45 mL) sugar
2 cups (500 mL) fresh, sliced strawberries
Plain yogurt or whipped cream
Confectioners' sugar

In an ovenproof skillet melt 2 tablespoons (30 mL) of butter in the oven at 425°F (220°C).

In a medium bowl beat the eggs until light. Gradually beat in the flour and buttermilk. Add the sugar and beat until well blended. Pour into the heated skillet and bake for 20 to 25 minutes until puffed and golden.

Place the sliced strawberries in the center and sprinkle with confectioners' sugar. Serve hot with sweetened yogurt or whipped cream.

The recipe may be doubled if desired.

Serves 4.

Strawberry Almond Crepes

DELICIOUS!

1 4-ounce (125 g) package cream cheese, softened
2 cups (500 mL) fresh strawberries, halved
¼ cup (50 mL) sliced almonds, toasted
4 tablespoons (60 mL) butter or margarine
1 cup (250 mL) red currant jelly
¼ cup (50 mL) Amaretto liqueur
¼ cup (50 mL) water

16 dessert crepes (purchased or use your favourite crepe recipe)

In a medium bowl whip the cream cheese with an electric mixer until fluffy. Spread over one side of each crepe and fold into quarters. Arrange the quartered crepes in a circle in a crepe pan. Place the strawberries in the center. In a small saucepan combine the butter, currant jelly, liqueur, and water, and heat and stir until bubbling. Pour the sauce over the crepes and strawberries, and heat together for 1 or 2 minutes. Serve on individual dessert plates. Top with whipped cream topping if desired and sprinkle with toasted almonds.
   Serves 8.

French Toast with Strawberries

4 eggs
2 cups (500 mL) buttermilk
2 tablespoons (30 mL) sugar
1 teaspoon (15 mL) ground cinnamon
French bread, thickly sliced
2 tablespoons (30 mL) butter or margarine

Strawberry Topping:
2 cups (500 mL) fresh strawberries
2 tablespoons (30 mL) butter
3 tablespoons (45 mL) sugar
¼ cup (50 mL) fresh orange juice

In a large bowl combine the eggs, buttermilk, sugar, and cinnamon, and beat or whisk until smooth and well blended. Soak each slice of bread completely in the egg mixture. Heat a large skillet over medium heat. Melt 2 tablespoons (30 mL) of butter in the pan and cook the slices until golden brown on both sides. Place in a warm oven until all of the slices are ready to serve.

Slice 1 cup (250 mL) of strawberries into halves. In a food processor or blender purée the remaining strawberries. In a clean skillet melt 1 tablespoon (15 mL) of butter over medium heat. Sauté the strawberry halves for 1 to 2 minutes just to warm. Do not over-cook. Remove from the skillet. Melt 1 tablespoon (15 mL) of butter in the skillet and stir in the sugar. Add the puréed strawberries and orange juice. Bring to a boil, stirring over medium heat for 1 minute. Pour the syrup over the sautéed strawberries.

Serve immediately over French toast.

Serves 6.

Fresh Strawberry Omelets

2 cups (500 mL) fresh strawberries
2 tablespoons (30 mL) sugar
½ teaspoon (2 mL) grated lemon peel
1 tablespoon (15 mL) fresh lemon juice
4 eggs
2 tablespoons (30 mL) butter or margarine

Wash and hull the strawberries. Slice into a bowl and sprinkle with sugar, lemon peel, and juice. Cover and let stand at room temperature for 15 minutes.

To prepare the omelets, in a medium bowl beat the eggs with a fork. In a skillet melt half the butter over medium-high heat. Pour half the egg mixture into the pan. With a pancake turner, gently push the cooked portions of egg at the edge into the center. Tilt the pan to ensure that the eggs are completely cooked. While the top is still creamy in texture, spread one half of the strawberry mixture over half the omelet. Fold. Keep warm. Repeat to make the second omelet.

Serves 2.

Baked Brie with Strawberries

ELEGANT FOR BRUNCH OR AS AN APPETIZER!

1 fresh round loaf bread, unsliced
1 small round Brie cheese
Fresh strawberries, sliced lengthwise

Slice off the top third of the bread. Cut and scoop out the center of the bread, leaving 1 inch of bread on the bottom and sides. Trim the Brie to fit and insert it into the loaf. Wrap in foil. Bake at 350°F (180°C) for 30 minutes until heated through.

Arrange sliced strawberries over the top. Cut into wedges and serve warm.
Serves 8.

Fresh Strawberries with Croissants and Almond Custard Sauce

A DELICIOUS WAY TO START A DAY!

Almond Custard Sauce:
3 eggs, lightly beaten
$\frac{1}{4}$ cup (50 mL) sugar
2 cups (500 mL) milk, scalded
1 tablespoon (15 mL) butter
$\frac{1}{2}$ teaspoon (2 mL) almond extract

4 fresh croissants
2 cups (500 mL) sliced fresh strawberries
$\frac{1}{2}$ cup (125 mL) sliced almonds, toasted

In the top of a double boiler over simmering water combine the eggs, sugar, and scalded milk. Cook, stirring constantly, until the mixture coats a metal spoon. Remove from the heat and stir in the butter and almond extract. Place the pan in cold water, and stir until cooled. Chill. Makes 2 cups (500 mL).

To serve, with a very sharp knife slice the croissants in half. On the lower half spread 2 tablespoons (30 mL) of cool custard and top with sliced strawberries. Replace the top half of the croissants. Bake at 300°F (150°C) for 6 minutes or until warm.

Warm the remaining almond custard sauce over low heat. Remove the croissants from the oven and place on serving plates. Sprinkle with toasted almonds. Serve with additional warm almond sauce.

Serves 4.

Strawberry Coffee Cake

1 15-ounce package (425 g) frozen strawberries, thawed
1 tablespoon (15 mL) cornstarch

2¼ cups (550 mL) all-purpose flour
¾ cup (175 mL) sugar
¾ cup (175 mL) margarine
1 teaspoon (5 mL) baking soda
¾ cup (175 mL) buttermilk
1 egg, lightly beaten

In a small bowl mix a little of the strawberry juice with the cornstarch and stir the mixture into the strawberries. Transfer the mixture to a saucepan. Cook and stir over low heat until thickened. Strain the mixture and discard the seeds. Set aside.

In a large bowl combine the flour and sugar. Cut in the margarine to fine crumbs. Reserve ½ cup (125 mL) of the crumb mixture. To the remaining crumb mixture add the baking soda. In a small bowl combine the buttermilk and lightly beaten egg and stir into the crumb mixture. Spread two-thirds of the mixture over the bottom and halfway up the sides of a deep 9-inch (22 cm) quiche or cake pan. Spread the strawberry mixture over the top. Spoon the remaining batter in mounds over the top of the strawberry mixture. Sprinkle the reserved crumbs on top. Bake at 350°F (180°C) for 40 minutes. Serve warm.

Serves 8.

Fresh Strawberry Coffee Cake

1 cup (250 mL) all-purpose flour
¼ cup (50 mL) sugar
2 teaspoons (10 mL) baking soda
½ cup (125 mL) buttermilk
1 egg
2 tablespoons (30 mL) melted margarine
1½ cups (375 mL) sliced fresh strawberries

Topping:
½ cup (125 mL) firmly packed brown sugar
½ cup (125 mL) all-purpose flour
½ teaspoon (5 mL) ground cinnamon
¼ cup (50 mL) margarine
¼ cup (50 mL) chopped pecans

In a large bowl combine 1 cup (250 mL) of flour, the sugar, and baking soda. Add the buttermilk, egg, and margarine, and beat until well blended. Spread the batter in a greased square cake pan. Place sliced berries over the top of the batter.

For the topping, in a small bowl combine the brown sugar, ½ cup (125 mL) of flour, and cinnamon. Cut in the margarine until crumbly. Stir in the pecans. Sprinkle the topping over the strawberries. Bake at 375°F (190°C) for 35 minutes.

Serve warm with whipped topping if desired.
Serves 9.

Strawberry Skillet Shortcake

A QUICK AND EASY DESSERT.

1 8-ounce (250 g) package buttermilk refrigerator biscuits
2 tablespoons (30 mL) butter
2 tablespoons (30 mL) firmly packed brown sugar
3 cups (750 mL) sliced fresh strawberries, sweetened to taste
1 tablespoon (15 mL) honey
1 cup (250 mL) yogurt

Separate the biscuits into halves. In a skillet melt the butter over low heat. Sprinkle in the brown sugar. Place the biscuits in the skillet and press down to flatten. Saute for 3 to 4 minutes on each side or until done. Divide the hot biscuits among serving plates, and top with sweetened berries.

In a small bowl stir the honey into the yogurt, and drizzle over the strawberries and biscuits.

Serves 4 to 6.

Fresh Strawberry Bread

A FLAVORFUL, LIGHT TEA BREAD.

1½ cups (375 mL) all-purpose flour
¾ cup (175 mL) sugar
1 teaspoon (5 mL) baking soda
1 teaspoon (5 mL) ground cinnamon
⅓ cup (75 mL) slivered almonds

¼ cup (50 mL) light cooking oil
¾ cup (175 mL) buttermilk
1 cup (250 mL) fresh strawberries, mashed
2 eggs, beaten

In a large bowl combine the flour, sugar, baking soda, cinnamon, and almonds. Make a well in the center of the flour mixture and add the oil, buttermilk, mashed strawberries, and eggs.

Stir well and pour into a greased loaf pan. Bake at 350°F (180°C) for 40 to 45 minutes. Slice. Serve warm with whipped topping, garnished with fresh whole strawberry. **Makes 1 loaf.**

Strawberry Pizza

A GREAT FAVOURITE FOR KIDS OF ALL AGES!

Crust:
2 cups (500 mL) all-purpose flour
1/4 cup (50 mL) sugar
1/2 teaspoon (2 mL) salt
4 teaspoons (20 mL) baking powder
1 cup (250 mL) margarine
1 egg, beaten
1/2 cup (125 mL) milk
1 teaspoon (5 mL) vanilla extract

Topping:
4 cups (1 L) fresh strawberries
1/3 cup (75 mL) red currant jelly
1 tablespoon (15 mL) cornstarch
1/3 cup (75 mL) apple juice

In a large bowl sift together the flour, sugar, salt, and baking powder. Cut in the margarine until crumbly. In a separate bowl combine the egg, milk, and vanilla, and stir quickly into the dry ingredients. Roll out on a floured board to fit a 12-inch (30 cm) pizza pan. Trim and flute the edge high enough to hold the strawberries. Prick the pastry with a fork. Bake at 400°F (200°C) for 10 to 15 minutes or until golden. Cool completely.

Wash and hull the strawberries. Cut in half lengthwise and arrange on cooled crust in circular fashion; flat side down. In a small saucepan melt the jelly over low heat. In a small bowl combine the cornstarch and apple juice. Stir into the melted jelly. Cook and stir until the mixture bubbles and thickens. Cook for 2 more minutes. Cool slightly. Spoon over the berries on the crust. Cover and chill until ready to serve.

To serve, cut into wedges. Garnish with whipped topping if desired.
Serves 10.

Fresh Strawberry Muffins

2 cups (625 mL) all-purpose flour
¾ cup (175 mL) sugar
1 teaspoon (5 mL) baking soda
1½ cups (375 mL) sliced strawberries
1 cup (250 mL) buttermilk
¼ cup (50 mL) margarine, melted
1 teaspoon (5 mL) almond extract
2 egg whites, stiffly beaten

Line a muffin pan with paper liners. In a large bowl combine the flour, sugar, and baking soda. Stir in the sliced strawberries. Make a well in the center. In a separate bowl combine the buttermilk, margarine, and almond extract. Pour into the well of the dry ingredients, and stir just to moisten. Fold in the stiffly beaten egg whites. Spoon the batter into the paper liners. Bake at 350°F (180°C) for 15 to 20 minutes.

Turn out on wire rack and cool.
Makes about 12 standard muffins.

Strawberry Pinwheel Sandwiches

GREAT FOR WEDDING SHOWERS AND SPECIAL OCCASIONS.

2 cups (500 mL) fresh strawberries
Sugar
1 8-ounce (250 g) package cream cheese, softened
3 tablespoons (45 mL) sour cream
1 loaf unsliced sandwich bread

Wash and hull the strawberries. In a bowl crush the strawberries slightly, sweeten to taste, and drain the juice. In a medium bowl beat the cream cheese and sour cream together until fluffy. Gently fold in the drained strawberries. Remove the crusts from the bread and slice lengthwise into 4 or 5 very thin slices. Spread each slice with the cream cheese mixture. Roll up, beginning at the narrow end. Wrap each roll in plastic wrap and chill. Slice into rounds to serve.

Makes 3 to 4 dozen pinwheel sandwiches.

Strawberry Cheese Spread

1 cup (250 g) cream cheese, softened
½ cup (125 g) vanilla-flavored yogurt
½ cup (125 g) puréed strawberries
2 tablespoons (30 mL) sugar

In a medium bowl blend all of the ingredients together. Chill.

Serve as spread for toast and muffins. May also be used as a dip for fresh fruit chunks. **Makes 2 cups.**

Salads

and

Main Courses

Strawberry and Spring Flower Salad with Lemon Honey Cream Dressing

2 cups (500 mL) fresh strawberries, halved

1 cup (250 mL) fresh, edible flowers (violets, pansies, rose petals, geranium petals, nasturtiums, apple and plum blossoms)

2 cups (500 mL) mixed young salad greens (lettuce, watercress, spinach)

Lemon Honey Cream Dressing:

½ cup (125 mL) heavy cream

1 tablespoon (15 mL) liquid honey

2 teaspoons (10 mL) fresh lemon juice

1 teaspoon (5 mL) lemon zest

Rinse the strawberries, flowers, and greens under cold water. Dry on a paper towel. Arrange the greens on salad plates. Place the strawberries over greens. Sprinkle the spring flowers over all. Chill until ready to serve.

In a blender combine the cream, honey, lemon juice, and zest. Blend well. Chill. Serve over salad.

Serves 4 to 6.

Strawberry Grenadine Salad

DELICIOUSLY REFRESHING!

1 3-ounce (85 g) package strawberry gelatin
1 cup (250 mL) boiling water
½ cup (125 mL) cold water
½ cup (125 mL) grenadine syrup
2 tablespoons (30 mL) lemon juice

1 cup (250 mL) sliced, fresh strawberries
⅓ cup (75 mL) flaked coconut

Dressing: (optional)
½ cup (125 mL) cream cheese, softened
¼ cup (50 mL) orange juice

In a large bowl dissolve the gelatin in boiling water. Add the cold water, grenadine, and lemon juice. Refrigerate until the consistency of unbeaten egg white.

Fold the strawberries and coconut into the gelatin mixture. Pour into individual salad molds and chill until set. To serve, unmold onto crisp lettuce leaves.

In a small bowl beat the cream cheese and orange juice together until smooth. Spoon over the salad.

Serves 4 to 6.

Strawberries and Bitter Greens

1 bunch dandelion greens
1 bunch watercress
1 cup (250 mL) sliced strawberries
Feta cheese, crumbled

Lemon Cream Dressing:
Lemon yogurt
Honey to taste

Wash and dry the greens. Arrange greens on 4 salad plates. Top with strawberries and feta. In a small bowl stir together lemon yogurt and honey to taste, and serve over the salad. **Serves 4.**

Strawberry and Shrimp Salad

2 cups (500 mL) quartered fresh strawberries
2 cups (500 mL) fresh shrimp, cooked
1 cup (250 mL) peeled and cubed kiwi
1 cup (250 mL) cubed papaya, cubed

Strawberry Cream Dressing:
$^1/_4$ cup (50 mL) strawberry purée
$^1/_4$ cup (50 mL) heavy cream
2 tablespoons (30 mL) rice wine vinegar
2 tablespoons (30 mL) olive oil
Honey

In a large bowl combine the strawberries, shrimp, kiwi, and papaya. Chill.

In a medium bowl combine the strawberry purée, cream, rice wine vinegar, and olive oil. Whisk together and let stand for 1 hour. Sweeten with honey if desired.

When ready to serve divide the salad among salad plates and serve with Strawberry Cream Dressing.

Serves 6.

Strawberry, Avocado, and Pecan Salad with Lime-Honey Dressing

AN ELEGANT, COLORFUL LUNCHEON SALAD OR STARTER FOR A DINNER PARTY.

3 cups (750 ml) avocado, cubed
2 cups (500 mL) strawberries, quartered
2 tablespoons (30 mL) fresh lime juice
1 cup (250 mL) pecans, toasted
Fresh lettuce leaves

Lime-Honey Dressing:
1/2 cup (125 mL) mayonnaise
2 tablespoons (30 mL) lime juice
1 tablespoon (15 mL) honey

In a salad bowl combine the avocado and strawberries. Sprinkle with 2 tablespoons (30 mL) of lime juice. Add the toasted pecans. Place lettuce leaves on serving plates and top with salad.

In a medium bowl combine the mayonnaise, 2 tablespoons (30 mL) of lime juice, and honey, and whisk until blended. Spoon over the salad.

Serves 6.

Strawberry Avocado Salad Mold

1 3-ounce (85 g) package lime gelatin
1 cup (250 mL) boiling water
⅔ cup (150 mL) cold water
2 tablespoons (30 mL) mayonnaise
1 ripe avocado, peeled and mashed
2 cups (500 mL) sliced strawberries
Whole strawberries

In a large bowl dissolve the gelatin in boiling water. Stir in the cold water and chill until partially set.

Add the mayonnaise and mashed avocado, and blend well. Fold in the sliced strawberries. Transfer to a lightly oiled gelatin mold. Chill for 3 to 4 hours until set. To serve, unmold onto a serving platter and garnish by arranging whole strawberries around the congealed salad.

Serves 6.

Strawberry Iceberg Salad

COOL AND ELEGANT!

1 3-ounce (85 g) package strawberry gelatin
1¾ cups (425 mL) boiling water
2 cups (500 mL) fresh strawberries, crushed

1 cup (250 mL) sour cream
½ cup (125 mL) cottage cheese
¼ cup (50 mL) mayonnaise
2 tablespoons (30 mL) confectioners' sugar
½ cup (125 mL) chopped walnuts

In a large bowl dissolve the gelatin in the boiling water. Chill until beginning to set.

Fold the crushed strawberries into the thickened gelatin. Divide evenly between 6 to 8 individual salad molds. Chill until completely firm.

In a medium bowl combine the sour cream, cottage cheese, mayonnaise, sugar, and walnuts. Mix well. Spread over the strawberry layer and freeze about 2 hours until firm. Unmold and serve on crisp lettuce leaves on salad plates.

Serves 6 to 8.

Strawberry and Spinach Salad

A COLORFUL, DELIGHTFUL SALAD.

2 bunches fresh spinach, washed and torn into pieces
1 tablespoon (15 mL) margarine
1½ cups (375 mL) washed, hulled, and quartered fresh strawberries
2 green onions, chopped
Lemon juice (optional)

In a large skillet sauté the spinach lightly in the margarine for 2 to 3 minutes or until the spinach is softened but not limp. Add the strawberries just to heat briefly (do not cook). Divide the salad among salad plates. Garnish with chopped green onion. Sprinkle with fresh lemon juice if desired.
Serves 6.

Strawberry and Cucumber Salad

2 cups (500 mL) peeled and thinly sliced English cucumber
Salt
1 cup (250 mL) washed, hulled, and sliced fresh strawberries

Dressing:
1 cup (250 mL) plain yogurt
1 garlic clove, finely minced
1 tablespoon (15 mL) lemon juice
1 tablespoon (15 mL) finely chopped chives

Place the sliced cucumber in a colander and sprinkle with salt. Let stand for 1 hour to allow the excess moisture to drain.

In a large bowl combine the cucumbers and strawberries, cover, and chill until ready to serve.

In a medium bowl mix the yogurt, garlic, lemon juice, and chives together. Spoon over individual servings of salad.

Serves 4.

Hot Crab Puffs with Curried Strawberries and Pecans

½ cup (125 mL) soft cream cheese
¼ cup (50 mL) mayonnaise
¼ cup (50 mL) minced sweet red bell pepper
¼ cup (50 mL) minced green onion
Salt and pepper to taste
1 cup (250 mL) fresh crab meat
8 slices Italian or other firmly textured bread

Curried Strawberries and Pecans:
2 cups (500 mL) strawberries, hulled
1 cup (250 mL) whole pecans
½ teaspoon (2 mL) curry powder
2 teaspoons (10 mL) sugar

In a large bowl combine the cream cheese and mayonnaise, and mix until smooth. Add the red pepper, onion, salt, pepper, and crab meat. Mix until well blended. Using a large fluted cookie cutter, cut the bread slices into rounds. Spread the crab mixture on bread rounds and place on a cookie sheet. Bake at 400°F (200°C) until golden brown.

Serve hot on a bed of chopped lettuce with Curried Strawberries and Pecans.

Cut the strawberries in half and set aside. In a lightly buttered skillet toast the pecans over medium heat. In a small bowl combine the curry powder and sugar, and sprinkle over the pecans. Sauté about 5 minutes until toasted. Add the strawberries to the pan only to warm the strawberries. Do not overheat. Serve with Crab Puffs.

Serves 4.

Blackened Snapper with Strawberries

$\frac{1}{2}$ teaspoon (2 mL) ground cinnamon
$\frac{1}{2}$ teaspoon (2 mL) garlic salt
$\frac{1}{4}$ teaspoon (1 mL) paprika
$\frac{1}{4}$ teaspoon (1 mL) oregano
$\frac{1}{4}$ teaspoon (1 mL) thyme
Salt and pepper to taste
2 fresh snapper fillets

1 tablespoon (30 mL) butter
$1\frac{1}{2}$ cups (375 mL) ripe strawberries, cut in half

In a small bowl combine the cinnamon, garlic, paprika, oregano, thyme, salt, and pepper. Season the fillets on both sides.

In a skillet melt the butter over medium heat. Sauté the strawberries briefly just to warm. Do not overheat. Remove from the pan and set aside. Increase the heat to high. Place the fillets in the skillet and cook over high heat until the fish is blackened but not burned. Turn over and cook the other side. Remove from the pan. Arrange spinach leaves on one side of each plate and place a fillet on top of each. Place sautéed strawberries on each plate.
Serves 2.

Chicken Fillet with Strawberry Peppercorn Cream Sauce

2 tablespoons (30 mL) all-purpose flour
Salt and pepper
2 chicken fillets
Butter and oil for frying

Strawberry and Peppercorn Cream Sauce:
½ cup (125 mL) thick cream
1 tablespoon (15 mL) green peppercorns
½ cup (125 mL) strawberries, quartered

In a shallow bowl mix the flour with salt and pepper. Coat the chicken fillets with the seasoned flour. In a skillet heat a small amount of butter and oil and cook the chicken over medium heat for 8 to 10 minutes on each side until golden brown. Remove from the skillet. Keep warm in the oven while preparing the sauce.

In a skillet heat the cream with the peppercorns over medium heat. Bring to a boil and cook for 2 minutes. Add the strawberries and cook for 1 minute. Serve the sauce over the chicken with fresh cooked asparagus spears.

Serves 2.

Chicken with Strawberries in Spicy Orange Sauce

4 chicken breasts
Salt and pepper
Butter and oil for frying

1 clove garlic, crushed
2 tablespoons (30 mL) chopped green onions
1 teaspoon (15 mL) grated fresh ginger
¼ teaspoon (1 mL) turmeric
¼ teaspoon (1 mL) curry powder
2 teaspoons (10 mL) firmly packed brown sugar
1 teaspoon (5 mL) cornstarch
½ cup (125 mL) fresh orange juice

1 cup (250 mL) strawberries, halved

Season the chicken with salt and pepper. In a skillet heat a small amount of butter and oil and cook the chicken breasts over medium heat until golden brown. Remove from the heat and keep warm.

In a skillet melt a small amount of butter and sauté the garlic and green onions over medium heat. Add the ginger, turmeric, curry powder, and brown sugar. In a cup mix the cornstarch with a little of the fresh orange juice and stir into the remaining orange juice. Add the juice to the skillet and stir until thickened. Stir in the strawberries and heat for 1 minute.

Serve over the chicken.

Serves 4.

Desserts

Strawberries Jubilee

A CELEBRATION DESSERT!

$\frac{1}{4}$ cup (50 mL) sugar
1 tablespoon (15 mL) butter
$\frac{1}{3}$ cup (150 mL) orange juice
2 cups (500 mL) small fresh strawberries, washed and hulled
$1\frac{1}{2}$ tablespoons (25 mL) Benedictine liqueur
1 teaspoon (5 mL) grated orange rind
1 tablespoon (15 mL) red currant jelly

3 tablespoons (45 mL) Benedictine liqueur

In a stainless steel skillet combine the sugar and butter and cook over low heat until the mixture caramelizes. Add the orange juice and cook, stirring well. Mash 4 strawberries and stir into the orange mixture. Add $1\frac{1}{2}$ tablespoons (25 mL) of Benedictine, the orange rind, and red currant jelly, and stir until well combined. Add the remaining strawberries and cook for 3 to 5 minutes, just to heat thoroughly. Stir constantly. Add 3 tablespoons (45 mL) of Benedictine liqueur and ignite with care. Shake the pan until the flame goes out.

Place a scoop of vanilla ice cream in each of 4 individual dessert glasses and divide the strawberry mixture evenly over each. Serve immediately.

Serves 4.

BOWL OF STRAWBERRIES

STRAWBERRY SUMMER SOUP, PAGE 16

STRAWBERRY ALMOND CREPES, PAGE 19

STRAWBERRY PIZZA, PAGE 28

Fresh Strawberry Muffins, page 29

STRAWBERRY AND SHRIMP SALAD, PAGE 37

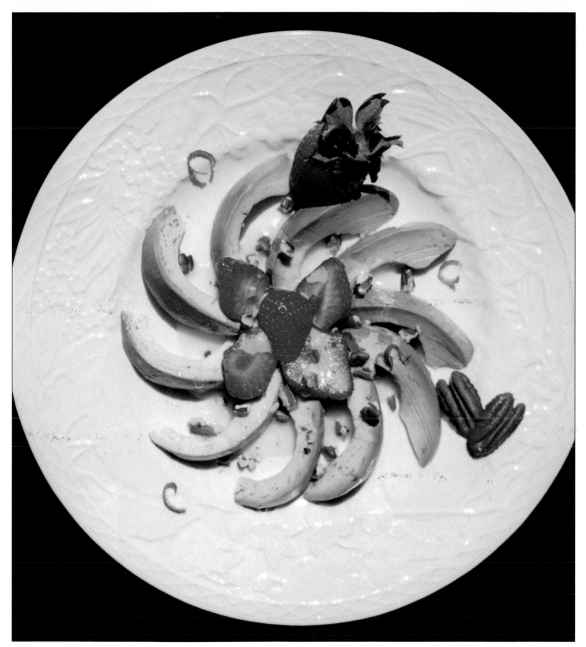

Strawberry, Avocado, and Pecan Salad with Lime and Honey Dressing, page 38

STRAWBERRY AND SPINACH SALAD, PAGE 41

HOT CRAB PUFFS WITH CURRIED STRAWBERRIES AND PECANS, PAGE 43

BLACKENED SNAPPER WITH STRAWBERRIES, PAGE 44

CHOCOLATE-DIPPED STRAWBERRIES, PAGE 65

STRAWBERRY MERINGUES, PAGE 66

STRAWBERRIES ROMANOV, PAGE 67

STRAWBERRY PARFAIT, PAGE 68

Strawberries and Hearts, page 71

STRAWBERRY ANGEL PARFAIT, PAGE 72

Chocolate-Dipped Strawberries

DECADENT AND ELEGANT!

6 ounces (170 g) semisweet chocolate
10 to 12 ripe strawberries with stems

Line a sheet pan with waxed paper. In the top of a double boiler over simmering water melt the chocolate.

Hold each strawberry by the stem and dip into the melted chocolate. Leave a little red showing for contrast. Place on the waxed paper and refrigerate.

Serves 2.

Strawberry Meringues

A DELIGHTFUL DESSERT.

4 egg whites
¼ teaspoon (1 mL) cream of tartar
1 cup (250 mL) sugar
½ teaspoon (2 mL) almond extract
4 cups (1 L) fresh ripe strawberries
Sweetened whipped cream or whipped topping

In a large bowl beat the egg whites and cream of tartar until soft peaks form. Slowly add the sugar and continue to beat until stiff and glossy. Add the almond extract. To make meringue shells, use ⅓ cup (75 mL) for each shell and shape with a spoon. Place on baking sheets lined with parchment or two layers of waxed paper. Bake at 275°F (135°C) for 1 hour. Turn the oven off and let cool to room temperature.

Wash and hull the strawberries. Save 12 whole strawberries and slice the remainder. Arrange in meringue shells in circular fashion. Top with whipped cream and a whole strawberry.

Variation: You may wish to make thin meringue wafers by tracing circles around a glass on a parchment paper, then filling each circle with a tablespoon of meringue and smoothing evenly. Bake at 225°F (110°C) for 30 minutes. Turn off the oven and let cool to room temperature. To assemble, cover one wafer with whipped cream. Place a layer of sliced strawberries over the top in a circular fashion, pointed ends facing outward. Top with a second meringue wafer. Place a small amount of whipped cream and a strawberry half on top of each.

Serves 12.

Strawberries Romanov

A CLASSIC RUSSIAN DESSERT!

4 cups (1 L) fresh strawberries
2 tablespoons (30 mL) confectioners' sugar
2 cups (500 mL) vanilla ice cream
1 cup (250 mL) whipping cream
1 tablespoon (15 mL) orange-flavored liqueur (optional)

Wash and hull the strawberries. Set aside 6 to 8 large berries for garnish. Slice the remaining berries into a serving bowl, sprinkle with confectioners' sugar, and chill.

Let the ice cream soften slightly. With a fork, work the liqueur into the softened ice cream.

Whip the cream until stiff, and fold into the ice cream mixture. Cover and freeze for about 1 hour.

To serve, place sliced strawberries in bottom of chilled sherbet glasses. Spoon ice cream mixture over top and garnish with a whole strawberry.

Serves 6 to 8.

Strawberry Parfait

2 cups (500 mL) fresh, ripe strawberries
1 3-ounce (85 g) package instant vanilla pudding mix
1 cup (250 mL) whipping cream or whipped topping
¼ cup (50 mL) flaked coconut

Wash and hull the strawberries. Reserve 4 to 6 whole strawberries, and slice the remainder. In a large bowl prepare the pudding according to the package directions. Cool to lukewarm.

In a separate bowl whip the cream and sweeten to taste. Spoon alternate layers of pudding and sliced strawberries into parfait glasses. Top with whipped cream and sprinkle with coconut. Garnish with whole berries. Chill until serving.

Serves 4 to 6.

Strawberries in Coconut Cream

4 cups (1 L) fresh strawberries
1 cup (250 mL) whipping cream
3 teaspoons (15 mL) confectioners' sugar
½ cup (125 mL) flaked coconut
½ teaspoon (2 mL) almond extract

Wash and hull the strawberries. Slice or leave whole if you prefer. In a bowl whip the cream, adding the sugar slowly, until stiff. Fold in half of the coconut and the almond extract. Place ⅓ cup (75 mL) of the mixture in each chilled sherbet glass, top with strawberries, and sprinkle with the remaining coconut.

Serves 6.

Strawberry Cream

1 15-ounce (425 g) package frozen strawberries, thawed
1 3-ounce (85 g) package strawberry gelatin
1 cup (250 mL) boiling water
1 cup (250 mL) whipping cream, whipped

Drain the strawberries, reserving the syrup. In a large bowl dissolve the gelatin in boiling water. Add enough cold water to the reserved syrup to make 1 cup (250 mL) of liquid, and stir into the gelatin mixture. Chill until partially set.

Beat with an electric mixer until foamy. Fold in the strawberries and whipped cream. Pour into individual molds and chill.

To serve, unmold onto dessert plates and top with additional whipped cream if desired. **Serves 6.**

Strawberry Delight

1 cup (250 mL) vanilla or graham cracker crumbs

$\frac{1}{2}$ cup (125 mL) margarine
1$\frac{1}{2}$ cups (375 mL) confectioners' sugar
2 egg yolks
$\frac{1}{2}$ teaspoons (2 mL) almond extract
2 egg whites, stiffly beaten

2 15-ounce (850 g) packages frozen strawberries, thawed and drained
$\frac{1}{2}$ cup (125 mL) chopped pecans
1 cup (250 mL) whipping cream, whipped

Spread half of the wafer crumbs on the bottom of an 8-inch square pan.

In a large bowl beat the margarine, confectioners' sugar, egg yolks, and almond extract together. Fold in the beaten egg whites. Spoon over the crumb base and spread evenly with a knife that has been dipped in hot water. Next layer with drained strawberries and sprinkle with pecans. Top with whipped cream and sprinkle with the remaining crumbs. Refrigerate for several hours before serving.

Serves 8 to 10.

Strawberries and Hearts

3 cups (750 mL) strawberries
1 sheet frozen puff pastry, thawed
2 tablespoons (30 mL) confectioners' sugar
½ teaspoon (2 mL) ground cinnamon

Preheat the oven according to the puff pastry package directions.

Roll the pastry out on a floured surface to ⅛-inch (0.25 cm) thickness. With a small heart-shaped cutter, cut out hearts and place on a cookie sheet. Sprinkle with confectioners' sugar and cinnamon. Bake for 10 to 15 minutes until puffed and golden brown.

Cool.

To serve, slice the strawberries lengthwise and arrange on individual dessert plates. Place the strawberries in a circular fashion with the pointed end of each strawberry toward the edge of the plate. Top with cinnamon hearts and serve. You may wish to garnish with whipped topping.

Serves 6 to 8.

Strawberry Angel Parfait

VERY SIMPLE AND DELICIOUS.

2 cups (500 mL) strawberries
½ cup (125 mL) sugar
2 cups (500 mL) whipping cream or whipped topping
Angel cake, cut into 1-inch pieces

Slice the strawberries into a bowl and sprinkle with half of the sugar. Set aside until the berries form juice. In a large bowl whip the cream, adding the remaining sugar slowly, until stiff. In parfait glasses alternate layers of berries and juice with layers of whipped cream and angel cake. Top with the remaining whipped cream and a whole berry.
Serves 4.

Strawberries Royale

AN ELEGANT TOUCH!

2 cups (500 mL) strawberries
2 tablespoons (2 mL) sugar
⅔ cup (150 mL) chilled Champagne
1 tablespoon (15 mL) Crème de Cassis

Wash and hull the strawberries. Place the strawberries in dessert glasses and sprinkle with sugar. Mix the Champagne and Crème de Cassis, pour over the strawberries, and serve.
Serves 2 to 4.

Peppered Strawberries

A UNIQUE AND REFRESHING DESSERT!

2 cups (500 mL) ripe strawberries
¼ cup (50 mL) sugar
½ teaspoon (2 mL) freshly ground black pepper
2 teaspoons (10 mL) grated orange rind
¼ cup (50 mL) fresh orange juice

Wash and hull the strawberries. Cut into halves and place in a serving bowl. Add the sugar, black pepper, orange rind, and juice. Cover and allow to stand at room temperature for 1 hour.

To serve, sprinkle with additional ground pepper if desired.
Serves 4.

Strawberries Mystère

AN OLD FRENCH RECIPE.

4 cups (1 L) fresh strawberries
¼ cup (50 mL) sugar
¼ cup (50 mL) red wine vinegar

Wash and hull the strawberries. Cut the small and medium berries in half, larger berries in quarters. Place in a large bowl and sprinkle with sugar and red wine vinegar. Let stand at room temperature for 30 minutes to marinate.

Serve in chilled wine glasses.

This may also be used as a topping for ice cream or cake.

Serves 6.

Strawberries and Chantilly Cream

4 cups (1 L) fresh strawberries
1 cup (250 mL) whipping cream
⅓ cup (75 mL) confectioners' sugar
½ cup (125 mL) sour cream
½ teaspoon (2 mL) vanilla extract
Grated white chocolate (optional)

Wash and hull the strawberries. In a large bowl whip the cream until stiff. Fold in the sugar, sour cream, and vanilla. Serve as a topping for the strawberries. Sprinkle with grated white chocolate.

Serves 6 generously.

Strawberry Zabaglione

A DELICIOUS FINALE TO A SPECIAL DINNER!

2 cups (500 mL) sliced strawberries
3 egg yolks
3 tablespoons (45 mL) sugar
2 tablespoons (30 mL) orange liqueur

Divide the sliced strawberries equally among 4 large wine glasses. In the top of a double boiler over simmering water combine the egg yolks and sugar. Beat with an electric hand mixer on high speed about 5 minutes until thick and double in volume. Gradually beat in the liqueur. Continue to beat until thickened and the mixture forms mounds. Spoon the warm zabaglione over the strawberries. Garnish each with a whole strawberry and serve immediately.

Serves 4.

Baked Strawberry Soufflé

8 egg whites
½ teaspoon (2 mL) cream of tartar
½ teaspoon (2 mL) salt
⅔ cup (150 mL) sugar
2 cups (500 mL) fresh strawberry purée

Preheat the oven to 350°F (180°C). Lightly grease a 3-quart (3 L) metal mold.

In a large mixing bowl beat the egg whites, cream of tartar, and salt with an electric mixer on high speed until soft peaks form. Gradually add the sugar, continually beating until stiff and glossy. Gently fold in half of the strawberry purée, just until blended. Fold in the remaining purée. Pour into the prepared mold. Set in a pan containing 1 inch (2.5 cm) of hot water. Bake for 45 to 50 minutes or until puffed and golden brown.

Cool about 30 minutes. To unmold, loosen the edge of the soufflé from the mold and invert onto a serving platter. Serve warm with sweetened whipped cream.

Serves 10.

Strawberry Soufflé

FOOD FOR THE GODS!!

2½ cups (625 mL) fresh strawberries (reserve 6 whole berries for garnish)
1 envelope unflavored gelatin
¼ cup (50 mL) cold water
4 eggs yolks
½ cup (125 mL) sugar
1 tablespoon (15 mL) lemon juice
4 egg whites
¼ cup (50 mL) sugar
1 cup (250 mL) whipping cream

Prepare a 4-cup (1 L) soufflé dish by making a collar of waxed paper or foil and securing firmly with tape.

In a blender purée the strawberries on high speed for 30 seconds.

In a small bowl soak the gelatin in cold water until softened.

In the top of a double boiler beat the egg yolks and ½ cup (125 mL) of sugar until lemon colored. Add the lemon juice and cook over simmering water, stirring constantly, for about 5 minutes until thickened. Add the gelatin and stir until dissolved. Cool. Stir in the strawberry purée. Add 1 or 2 drops of red food coloring to tint a delicate pink.

In a large bowl beat the egg whites until soft peaks form. Slowly add ¼ cup (50 mL) of sugar and continue to beat until stiff.

In a separate bowl whip the cream until soft peaks form. Fold the meringue and whipped cream into the strawberry mixture. Pour into the prepared soufflé dish. Chill for about 3 hours until firm.

Remove the collar. Decorate individual servings with a whole berry.
Serves 6.

Meringue Soufflé
with Strawberries and Custard Sauce

4 egg whites
½ cup (125 mL) sugar
1 teaspoon (5 mL) almond extract
2 cups (500 mL) fresh strawberries
Confectioners' sugar

Custard Sauce:
2 cups (500 mL) milk
3 whole eggs or 6 egg yolks
¼ cup (50 mL) sugar
½ teaspoon (2 mL) vanilla extract

In a large bowl beat the egg whites until soft peaks form. Gradually add the sugar, continually beating until stiff and glossy. Flavor with almond extract. Spoon into 4 buttered individual soufflé dishes and place in a shallow pan of boiling water. Bake at 350°F (180°C) for 18 to 20 minutes. Cool slightly and invert onto dessert plates.

In the top of a double boiler over simmering water scald the milk. In a medium bowl lightly beat the eggs and stir in the sugar. Add a little scalded milk to the egg mixture and stir. Return to the scalded milk in the double boiler, stirring constantly until the custard is slightly thickened and coats the back of a metal spoon. Remove from the heat, and stir in the vanilla. Cool.

Spoon Custard Sauce around each soufflé. Top with strawberries and dust with sugar.
Serves 4.

Strawberries in Chocolate

DECADENT!

Whole fresh, ripe strawberries, small to medium in size
10 ounces (300 g) semisweet chocolate
¼ cup (50 mL) light cream
2 cups (500 mL) whipping cream
2 tablespoons (30 mL) sugar

Rinse and hull the strawberries. In a heavy-bottomed saucepan melt the chocolate with the cream over low heat. Stir until smooth and blended. In the bowl of an electric mixer beat the cream with the sugar until stiff.

Place 2 tablespoons (30 mL) of melted chocolate in the bottom of 4 long-stemmed dessert or champagne glasses. Top with strawberries. Spoon generous portions of sweetened whipped cream over all. Serve immediately.

Serves 4.

Strawberries and White Chocolate Mousse

HEAVENLY!

8 ounces (250 g) white chocolate
2¼ cups (550 mL) heavy cream
⅓ cup (75 mL) confectioners' sugar
1 teaspoon (5 mL) vanilla extract

4 cups (1 L) sliced fresh strawberries
2 tablespoons (30 mL) liqueur (orange or almond flavored).

In the top of a double boiler over gently simmering water melt the chocolate. Blend in the cream and sugar, allowing to heat slightly until completely blended and the sugar is dissolved. Add the vanilla. Chill thoroughly for several hours.

Beat the mousse with an electric mixer on high speed until light and fluffy. Do not overbeat. In a large bowl combine the sliced strawberries and liqueur and serve over the mousse on dessert plates.

Serves 8.

Strawberry Mousse

4 cups (1 L) fresh strawberries
¾ cup (175 mL) sugar
½ cup (125 mL) water
2 envelopes unflavored gelatin
⅓ cup (75 mL) cold water
2 cups (500 mL) heavy cream, whipped

Slice and mash or purée 3 cups (750 mL) of the strawberries. Reserve 1 cup (250 mL) of berries for garnish.

In a saucepan combine the sugar and ½ cup (125 mL) of water, and bring to a full boil. Cook for 2 minutes. Remove the pan from the heat.

In a small bowl soak the gelatin in the cold water. Stir into the hot syrup until dissolved. Cool. Blend the mashed strawberries into the cool syrup mixture. Chill until partially set.

Fold the whipped cream into the chilled gelatin mixture. Pour into an 8-cup (2 L) dessert bowl or mold. Cover with plastic wrap and chill for 3 to 4 hours until completely set.

To serve, unmold the mousse onto a serving platter. Garnish with the reserved strawberries.

Serves 8.

Strawberry Chocolate Mousse

2 1-ounce squares semisweet chocolate
3 tablespoons (45 mL) hot water
6 ounces (170 g) light cream cheese
1 tablespoon (15 mL) confectioners' sugar
1/4 cup (50 mL) milk
2 cups (500 mL) whipping cream
2 tablespoons (30 mL) sugar
2 cups (500 mL) sliced strawberries

In a saucepan with a heavy bottom or in a double boiler melt the chocolate in the hot water over low heat, stirring until smooth. Remove from the heat and cool.

In a medium bowl beat half of the cream cheese with confectioners sugar until well blended. Add the milk and beat until blended. In a separate bowl whip the cream until it begins to thicken. Add the sugar and continue to beat until stiff. Do not overbeat. Mix 2 cups (500 mL) of the whipped cream into the creamed cheese mixture, then mix in 1 cup (250 mL) of the sliced strawberries. Transfer to a soufflé dish or dessert bowl. Chill.

Beat the remaining cream cheese until fluffy. Blend in the melted chocolate and remaining whipped cream. Cover the chilled strawberry-cream cheese mixture with the chocolate mixture. Chill completely for about 4 hours. Serve with the remaining sliced strawberries and additional whipped cream if desired.

Serves 6 to 8.

Snow Pudding with Strawberries in Crimson Raspberry Sauce

Snow Pudding:
1 cup (250 mL) sugar
1 envelope unflavored gelatin
1¼ cups (300 mL) milk
1 teaspoon (5 mL) almond extract
1¼ cups (300 mL) flaked coconut
2 cups (500 mL) whipping cream, whipped

Crimson Raspberry Sauce:
1 15-ounce (425 g) package frozen raspberries, thawed
2 teaspoons (10 mL) cornstarch
½ cup (125 mL) red currant jelly
2 cups (500 mL) sliced fresh strawberries

For the pudding, in a saucepan combine the sugar, gelatin, and milk. Heat over medium heat, stirring until the sugar and gelatin are dissolved. Cool and stir in the almond flavoring. Chill until partially set.

Fold in the coconut and then the whipped cream. Pour into a 1½-quart (1.5L) mold. Chill for about 4 hours until set.

For the sauce, in a saucepan crush the thawed raspberries. Stir in the cornstarch and jelly. Bring to a boil, stirring constantly until the mixture is slightly thickened and clear. Strain through a sieve and discard the seeds. Chill. Add the sliced strawberries and chill until ready to serve. Serve over snow pudding.

Serves 8.

Strawberry Trifle

A TRADITIONAL RECIPE.

Custard:
4 egg yolks
3 tablespoons (45 mL) sugar
1 cup (250 mL) light cream
1 cup (250 mL) milk
2 teaspoons (10 mL) vanilla extract

Trifle Assembly:
2 sponge cake layers
½ cup (125 mL) sherry
½ cup (125 mL) strawberry jam
1 cup (250 mL) sliced strawberries
1 cup (250 mL) whipping cream, whipped
Fresh whole strawberries

In double boiler beat egg yolks and sugar together. Stir in the light cream, milk, and vanilla. Cook over boiling water, stirring constantly, for about 10 to 15 minutes until the custard is thickened. Chill completely.

In a large serving bowl place one cake layer. Sprinkle half of the sherry over the cake. Spread with half of the strawberry jam. Pour half of the cooled custard over the top. Cover with sliced strawberries. Repeat the layers. Top with whipped cream and garnish with whole strawberries. Chill.

Serves 10.

Strawberries and Lime Custard

REFRESHING AND COOL.

1 cup (250 mL) light cream
Peel of $^1/_2$ lime
3 egg yolks
$^1/_2$ cup (125 mL) sugar
$^1/_2$ cup (125 mL) heavy cream
3 cups (750 mL) strawberries
2 tablespoons (30 mL) sugar or sweetener to taste
$^1/_4$ cup (50 mL) fresh lime juice

In a saucepan scald the light cream with the lime peel. In a large bowl beat the egg yolks with $^1/_2$ cup (125 mL) of sugar until thick and light. Stir in the scalded cream in a slow stream. Transfer to a saucepan and cook over medium heat, stirring constantly, just until the mixture coats a spoon. Strain the custard into a bowl and discard the lime peel. Cover and cool.

In a medium bowl lightly beat the heavy cream. Fold the whipped cream into the cooled custard. Chill.

Slice the strawberries in quarters. Add 2 tablespoons (30 mL) of sugar and the lime juice. Cover and chill.

Serve over chilled the custard.

Serves 4 to 6.

Strawberry Angel Trifle

Custard:
2 cups (500 mL) milk

3 eggs

¼ cup (50 mL) sugar

½ teaspoon (2 mL) vanilla extract

Trifle Assembly:
1 angel cake

¼ cup (50 mL) strawberry jam

2 tablespoons (30 mL) Amaretto liqueur

3 cups (750 mL) strawberries, sliced

2 cups (500 mL) custard (see above)

2 egg whites

1 cup (250 mL) whipping cream

2 tablespoons (30 mL) sugar

¼ cup (50 mL) sliced almonds, toasted

Whole strawberries for garnish

In the top of a double boiler over simmering water scald the milk. In a medium bowl beat the eggs and ¼ cup (50 mL) of sugar together. Add a little of the scalded milk to the egg mixture, then return the egg mixture to the double boiler. Cook over simmering water until the custard coats a metal spoon and is slightly thickened. Remove from the heat and stir in the vanilla. Chill completely.

Break the angel cake into pieces. Place half of the cake pieces in the bottom of a large glass serving bowl.

In a small bowl mix the jam and liqueur together. Drizzle half of the mixture over the cake pieces. Cover with half the strawberries. Spoon half of the custard over the strawberries.

Beat the egg whites until very stiff. In a separate bowl whip the cream with 2 tablespoons (30 mL) of the sugar. Fold in the egg whites. Cover the custard with a layer of egg white-cream mixture. Repeat the layers, ending with the cream layer. Garnish with toasted almonds and whole berries. Refrigerate. Serve within 3 hours.

Serves 10 to 12.

Strawberry Clouds

A DELICATE, CLOUD-LIKE DESSERT.

1½ cups (375 mL) sugar
½ cup (125 mL) water
3 egg whites, stiffly beaten
1 teaspoon (5 mL) almond extract
12 to 15 fresh strawberries, washed and hulled

Preheat the oven to 450°F (230°C). In a saucepan combine the sugar and water. Bring to a boil and cook without stirring to the soft ball stage, 234°F (110°C) on a candy thermometer.

Pour the hot syrup into the stiffly beaten egg whites, beating constantly until the mixture is very light and thick. Beat in the almond extract. Spoon a heaping tablespoon of meringue onto a lightly buttered baking sheet for each strawberry. Top each with a strawberry and cover with meringue, sealing the strawberry inside. Bake for 3 minutes or until a light golden brown. Watch very carefully. Cool and serve.

Serves 6.

Strawberries on a Cloud

Crust:
1½ cups (375 mL) all-purpose flour
¼ cup (50 mL) ground almonds
¼ cup (50 mL) sugar
⅓ cup (75 mL) margarine

Topping:
1 cup (8 oz.) cream cheese, softened
¾ cup (175 mL) confectioners' sugar
1 teaspoon (5 mL) vanilla or almond extract
2 cups (500 mL) whipping cream
3 cups (750 mL) miniature marshmallows
1 cup (250 mL) sliced strawberries
Whole strawberries

Preheat the oven to 400°F (205°C). In a large bowl combine the flour, ground almonds, and sugar. Cut in the margarine with a pastry blender. Press into a 9 x 13-inch (22 x 33 cm) pan. Bake for 10 to 12 minutes. Cool completely.

In a large bowl beat the cream cheese, confectioners' sugar, and vanilla or almond extract until smooth. In a separate bowl beat the chilled whipping cream until stiff peaks form. Fold the whipped cream into the cream cheese mixture. Fold in the marshmallows and then the sliced strawberries. Spread over the crust. Cover with plastic wrap and chill for 2 to 3 hours.

Top each serving with a whole strawberry.
Serves 8.

Strawberry Mallow Creme Dessert

Crust:
1½ cups (375 mL) graham cracker crumbs
¼ cup (50 mL) margarine, melted
3 tablespoons (45 mL) sugar

Filling:
½ cup (125 mL) milk
1 cup (250 mL) marshmallow creme
1½ cups (375 mL) whipped topping, thawed
3 cups (750 mL) sliced strawberries
Halved strawberries for garnish

In a medium bowl combine the graham crumbs, sugar, and margarine. Press into a 9-inch square (22 x 22 cm) baking pan. Chill.

In a saucepan combine the milk and marshmallow creme, and heat and stir over low heat until well blended. Cool to room temperature.

Place the whipped topping in a large bowl. Fold in the cooled marshmallow mixture and sliced strawberries. Spread the mixture over the chilled crust. Chill for 2 hours before serving.

Cut into squares and garnish with halved berries.
Serves 6.

Strawberry Coeur à la Crème

1 8-ounce (250 g) package light cream cheese, softened
1 cup (250 mL) whipping cream
¼ cup (50 mL) sugar
2 egg whites, beaten stiff

Strawberry Sauce:
2 cups (500 mL) strawberries
2 tablespoons (30 mL) sugar
1 tablespoon (15 mL) lemon juice
1 tablespoon (15 mL) strawberry liqueur

In a large bowl beat the cream cheese until fluffy. In a separate bowl whip the cream and ¼ cup (50 mL) of sugar until soft peaks form. Fold the whipped cream into the cream cheese until well blended. Fold in the stiffly beaten egg whites. Place the mixture in a fine wire sieve and place over a bowl to drain overnight in the refrigerator.

The next day divide the mixture into 4 to 6 individual heart-shaped molds lined with damp cheesecloth. Chill for several hours until firm.

In a blender purée the strawberries on low speed briefly. Stir in 2 tablespoons (30 mL) of sugar, the lemon juice, and liqueur. Cover and chill.

To serve, unmold the coeur à la crème onto dessert plates and garnish with Strawberry Sauce or sweetened, sliced strawberries.

Serves 4 to 6.

Strawberry Cheesecake Cups

INDIVIDUAL NO-BAKE CHEESECAKE CUPS.

1½ cups (375 mL) graham cracker crumbs
¼ cup (50 mL) margarine
2 egg yolks, beaten
½ cup (125 mL) milk
1 cup (250 mL) sugar
2 envelopes unflavored gelatin
½ cup (125 mL) cold water
1 teaspoon (5 mL) vanilla extract
2 8-ounce (250 g) packages light cream cheese, softened
2 egg whites, stiffly beaten
1 cup (250 mL) whipping cream, whipped
1 cup (250 ml) sliced strawberries

Line a muffin pan with paper baking cups. In a medium bowl mix the wafer crumbs and margarine together. Reserve a small amount for topping. In a saucepan combine the egg yolks, milk, and sugar; and cook over low heat, stirring until thickened.

In a small bowl soften the gelatin in cold water. Stir into the hot mixture. Add the vanilla. Cool.

Fold the mixture into the softened cream cheese. Fold in the stiffly beaten egg whites, then the whipped cream. Fold in the sliced strawberries. Spoon into the prepared baking cups. Sprinkle with the remaining crumbs. Chill until ready to serve.
Serves 12.

Strawberry Mille-Feuille

ALMOND PASTRY LAYERED WITH STRAWBERRIES AND WHIPPED CREAM.

Almond Pastry:
¾ cup (175 mL) sugar
½ cup (125 mL) ground almonds
¼ cup (50 mL) all-purpose flour
¼ cup (50 mL) butter, melted
¼ cup (50 mL) orange juice
2 teaspoons (10 mL) orange zest

1 cup (250 mL) whipping cream
2 tablespoons (30 mL) sugar

1 pint (500 mL) fresh strawberries, washed and sliced
Whole strawberries for garnish

In a medium bowl combine ¾ cup (175 mL) of sugar, the ground almonds, and flour. Stir in the melted butter, orange juice, and orange zest until well mixed. Drop the dough by teaspoons onto lightly oiled baking sheets to make 12 pastries. Flatten each with a floured fork. Bake at 400°F (200°C) for 6 minutes until golden. Allow to cool for 1 to 2 minutes on the baking sheet. Remove from the baking sheet to cool completely.

In a small bowl whip the cream and 2 tablespoons (30 mL) of sugar until soft peaks form.

To serve, on each of 4 dessert plates place an almond pastry. Cover each pastry with whipped cream and a layer of sliced strawberries. Add a dollop of whipped cream. Repeat the layering of pastry, whipped cream, and strawberries to make three layers for each mille-feuille. Top each with a whole strawberry.

Serves 4.

Chocolate Strawberry Puffs

6 individual puff pastry shells
1 cup (250 mL) dark chocolate chips, melted
Fresh strawberries, washed, hulled, and sliced in half lengthwise.
3 tablespoons (45 mL) red currant jelly
1½ tablespoons (25 mL) Amaretto liqueur, or water

Bake the puff pastry shells according to the package directions.

Spoon melted chocolate into the pastries. Cool slightly, about 15 minutes. Arrange the strawberries, cut side down, on top of the chocolate while it is still soft. Chill until the chocolate is set. In a saucepan heat the jelly and liqueur over low heat until the jelly is melted and smooth. Brush the strawberries with jelly glaze. Serve at room temperature.

Garnish with whipped topping if desired.

Serves 6.

Strawberry Cream Puffs

Cream Puff Dough:
1/4 cup (50 mL) butter or margarine
1/2 cup (125 mL) water
1/2 cup (125 mL) all-purpose flour
2 large eggs

Filling:
1 cup (250 mL) whipping cream
1 tablespoon (15 mL) confectioners' sugar
1 teaspoon (5 mL) vanilla extract
Sweetened sliced strawberries

Preheat the oven to 400°F (200°C). In a small saucepan bring the butter and water to a boil. Remove the pan from the heat and beat in the flour with a wooden spoon. Return to the heat and continue to beat until the mixture leaves the sides of the pan. Remove from the heat and beat in the eggs one at a time, beating until the dough is shiny and smooth. Drop the dough in mounds of about 1 tablespoon (15 mL) each onto an ungreased cookie sheet. Bake about 20 minutes until puffed and golden. Cool.

In a large bowl beat the cream with the sugar until stiff. Stir in the vanilla. Cut the tops off the puffs and fill with whipped cream and sweetened sliced strawberries. Replace the tops and sprinkle with confectioners' sugar. Chill until ready to serve.

Serves 6.

Strawberries in Puff Pastry

Puff Pastry:
*Frozen puff pastry sheet, thawed (you will need enough pastry
to make six 5 x 3-inch puffs)
1 egg white, beaten
Sugar*

Whipped Cream Filling:
*1½ cups (375 mL) whipping cream
2 tablespoons (30 mL) sugar
½ teaspoon (2 mL) almond extract*

Strawberry Filling:
*2 cups (500 mL) fresh strawberries
2 tablespoons (30 mL) sugar*

Preheat the oven according to the puff pastry package instructions. Cut the thawed pastry sheet into six 5 x 3-inch (12 x 7 cm) rectangles. Arrange the pastries on a cookie sheet. Brush each pastry with beaten egg white and sprinkle with sugar. Bake according to the package instructions until puffed and golden brown. When cool, split the pastries in half horizontally.

In a large bowl whip the cream until soft peaks form. Add 2 tablespoons (30 mL) of sugar slowly and beat until stiff. Add the almond extract.

Wash, hull, and slice the strawberries. Reserve six whole strawberries for garnish. Sprinkle the sliced strawberries with 2 tablespoons (30 mL) of sugar.

Just before serving fill the bottom of each pastry with whipped cream and top with sliced strawberries. Place the top half of the pastry puff on top. Garnish with additional whipped cream and a whole strawberry.

Serves 6.

Strawberry Cheesecake Royale

Crust:
Margarine
½ cup (125 mL) slivered almonds
1 cup (250 mL) vanilla wafer crumbs
⅓ cup (75 mL) margarine, melted

Cream Cheese Filling:
4 cups (1L) fresh strawberries, hulled
3 8-ounce (250g) packages cream cheese, softened
1 cup (250 mL) sugar
¾ cup (175 mL) fresh orange juice
2 envelopes unflavored gelatin
½ cup (125 mL) cold water
1 cup (250 mL) whipping cream, whipped

For the crust, lightly oil the bottom and sides of a 9-inch (22 cm) springform pan and line the sides with waxed paper. In a skillet melt a small amount of margarine and lightly toast the almonds. In a medium bowl combine the wafer crumbs, toasted almonds, and ⅓ cup (75 mL) of margarine. Mix well. Press into the pan. Chill.

For the filling, crush enough strawberries to make 2 cups (500 mL). Reserve the remaining berries for garnish. In a large bowl beat the cream cheese, sugar, and orange juice together with an electric mixer until smooth. Stir in the crushed strawberries. In a small saucepan soften the gelatin in cold water. Place the pan over low heat and cook, stirring constantly, until the gelatin is dissolved. Blend thoroughly into the cream cheese and strawberry mixture. Chill just until the mixture is thickened, not set.

Fold in the whipped cream. Pour into the prepared crust, and chill for several hours.

When ready to serve, remove the sides of the springform pan and the waxed paper. Decorate with additional whipped cream and whole strawberries.

Serves 12.

Strawberry Pavlova

AN ELEGANT DESSERT!

4 egg whites
1 cup (250 mL) sugar
2 teaspoons (10 mL) cornstarch
2 teaspoons (10 mL) vinegar
1 teaspoon (5 mL) almond extract
4 cups (1 L) fresh strawberries
2 cups (500 L) whipping cream

In a large bowl beat the egg whites with an electric mixer until stiff. Beat in the sugar gradually, beating until very stiff. Add the cornstarch, vinegar, and almond extract, beating well.

Lightly oil an attractive pie plate and pile the meringue so that it covers the bottom and sides in the form of a crust. Bake at 300°F (150°C) for 1 hour and 15 minutes.

Cool completely. The meringue will be crisp on the outside and soft inside. When ready to serve, slice the strawberries, reserving a few whole ones for garnish.

In a large bowl whip the cream until stiff. Place sliced strawberries in the meringue shell, and top with whipped cream and whole strawberries.

Serves 6.

Strawberry Angel Loaf

AN ULTRA LIGHT DESSERT!

Angel loaf:
½ cup (125 mL) sifted cake flour
¼ cup (50 mL) sugar
5 egg whites
½ teaspoon (1 mL) cream of tartar
1 teaspoon (5 mL) almond extract
½ cup (125 mL) sugar

2 cups (500 mL) heavy cream
2 tablespoons (30 mL) sugar
1 teaspoon (5 mL) almond extract

2 cups (500 mL) sliced strawberries

In a small bowl sift together the flour and ¼ cup (50 mL) of sugar. Set aside.

In a large bowl beat the egg whites and cream of tartar until soft peaks form. Add 1 teaspoon (5 mL) of almond extract. Add ½ cup (125 mL) of sugar slowly, beating constantly until stiff and glossy. Sift one-third of the flour mixture over the egg whites and fold in quickly. Repeat folding with the remaining flour in two additions. Transfer the angel batter to an ungreased glass loaf pan. Bake at 375°F (190°C) for 25 minutes.

Invert the pan and allow to cool. Remove the cake carefully from the pan and cut into 3 layers.

To assemble the cake, whip the heavy cream until soft peaks form. Add 2 tablespoons (30 mL) of sugar and beat until stiff. Fold in 1 teaspoon (5 mL) of almond extract. Spread a layer of the flavored whipped cream on the bottom layer of cake. Add a layer of sliced strawberries. Repeat with the second layer. Place the third layer on top and frost the top and sides of the loaf with the remaining whipped cream. Garnish with strawberries. Chill until ready to serve.

Note: Whipped topping may be used in place of whipped cream if preferred.
Serves 6.

Strawberry Shortcake and Devon Cream

Shortcake:
6 tablespoons (90 mL) soft butter
¼ cup (50 mL) sugar
½ cup (125 mL) light sour cream
1 egg
1½ cups (375 mL) all-purpose flour
2 teaspoons (10 mL) baking powder
½ teaspoon (2 mL) salt
¼ teaspoon (1 mL) baking soda

4 cups (1 L) ripe strawberries

Devon-Style Cream:
1 cup (250 mL) whipping cream
2 tablespoons (30 mL) light brown sugar
½ cup (125 mL) sour cream

In a large bowl cream the butter and sugar. Beat in the light sour cream and egg. In a separate bowl combine the flour, baking powder, salt, and baking soda, and stir into the butter mixture. Turn out onto a floured board and pat into a circle ½ inch (1 cm) thick. With a medium-size round cutter, cut 8 rounds of dough and place on a greased baking sheet. Bake at 400°F (200°C) for 12 minutes or until golden. Cool on a wire rack. Split the shortcakes in half horizontally.

Wash, hull, and slice the strawberries. Sweeten if desired.

In a large bowl whip the cream until stiff, adding the light brown sugar slowly. Fold in the sour cream. Chill for several hours.

To serve, fill the shortcakes with cream and strawberries. Top with additional cream and a whole strawberry.

Note: Devon Cream may be purchased at most food stores.

Serves 8.

Strawberry Angel Shortcake

LIGHT AND DELICIOUS!

3 cups (750 mL) plain yogurt
3 tablespoons (45 mL) sugar
½ teaspoon (2 mL) almond extract
1 angel cake
2 cups (500 mL) fresh sliced strawberries
Confectioners' sugar

Place the yogurt in a cheesecloth-lined sieve and place over a bowl. Cover and allow to drain overnight.

Discard the liquid. In a medium bowl blend the drained yogurt with the sugar and almond extract. Cut the angel cake into 3 layers. Spread the bottom layer with with half of the yogurt mixture. Arrange half of the strawberries over the yogurt. Place the second cake layer over the top and cover with the remaining yogurt mixture. Cover with the remaining strawberries. Add the third cake layer and dust with confectioners' sugar.

Serves 10 to 12.

Strawberry Whipped Cream Cake

A VERY SPECIAL CAKE.

2½ cups (625 mL) sugar
1 teaspoon (5 mL) salt
1 cups (250 mL) egg yolks
¼ cup (50 mL) egg whites
2 cups (500 mL) sifted cake flour
1 teaspoon (5 mL) baking powder
1 cup (250 mL) milk
½ cup (125 mL) butter
1 teaspoon (5 mL) vanilla extract

Strawberry Topping:
4 cups (1 L) whipping cream
1 tablespoon (15 mL) strawberry liqueur
2 cups (500 mL) chopped strawberries, sweetened
Whole strawberries for garnish

Grease three 8-inch (20 cm) round cake pans or two 10-inch (25 cm) springform pans. In the bowl of an electric mixer combine the sugar, salt, egg yolks, and egg whites. Beat at medium speed for 20 minutes or until stiff peaks form.

In a large bowl sift the flour and measure 2 cups. Add the baking powder and set aside.

In a small saucepan combine the milk and butter, and heat on low just until the butter is melted. Do not boil. Add the vanilla. Fold into the sugar-yolk mixture alternately with the dry ingredients. Blend well only, do not beat. Pour the batter into the prepared cake pans. Bake at 375°F (190°C) for about 15 to 20 minutes until golden.

In a large bowl whip the whipping cream. Blend the strawberry liqueur into the whipped cream. In a separate bowl mix chopped strawberries into half of the whipped cream and spread between the cake layers. Cover the cake with remaining whipped cream and garnish with whole berries. Refrigerate until ready to serve.

Note: This is a very delicate, fine-textured sponge cake that is quite wonderful.

Serves 8 to 10.

Strawberry Tiramisu

SIMPLY DELICIOUS!

3 eggs, separated
¾ cup (175 mL) confectioners' sugar
1½ cups (375 mL) light cream cheese, softened
¼ cup (50 mL) Amaretto liqueur
1 package ladyfinger cookies
2 cups (500 mL) strawberries, sliced
Grated bittersweet chocolate

In a large bowl beat the egg yolks, adding the confectioners' sugar gradually, until thick. Beat in the softened cream cheese and Amaretto until well blended. In a separate bowl beat the egg whites until stiff peaks form. Fold into the cream cheese mixture.

Line the bottom of a square glass baking dish with ladyfingers. Spread two-thirds of the cream cheese mixture over the ladyfingers. Cover with sliced strawberries. Add another layer of ladyfingers over the strawberries. Spread the remaining cream cheese mixture over the top. Sprinkle with grated chocolate. Cover and chill at least 4 hours. To serve, cut into squares and top each with a whole strawberry.

Serves 6.

Strawberry Angel Roll

LUSCIOUS AND LIGHT.

1 package angel food cake mix
1 cup (250 mL) whipping cream
2 tablespoons (30 mL) confectioners' sugar
2 cups (500 mL) fresh strawberries, sliced
Confectioners' sugar

Prepare the angel food cake batter according to the package directions. Spread half of the batter in waxed paper lined jelly roll pan. Bake at 375°F (190°C) for 20 minutes. Cool and remove the waxed paper.

Use the remaining batter for a second jelly roll or bake in a loaf pan for 25 minutes. In a large bowl whip the cream and sweeten with sugar. Spread on the cooled cake. Cover with sliced strawberries. Roll up jelly roll fashion. Wrap in plastic wrap and refrigerate for at least 2 hours.

To serve, sprinkle with confectioners' sugar and slice. Garnish each serving with a whole berry.

Serves 8 to 10.

Strawberry Almond Torte

Almond Filling:
1 cup (250 mL) almond paste

1 egg

Crust:
2 cups (500 mL) all-purpose flour

1/4 cup (50 mL) sugar

3/4 cup (175 mL) margarine

1 egg

1/2 teaspoon (2 mL) almond extract

Whipped Cream Topping:
2 cups (500 mL) whipping cream

2 tablespoons (30 mL) sugar

1 teaspoon (5 mL) almond extract

4 cups (1L) fresh strawberries

In a small bowl beat the almond paste and egg together.

In a large bowl combine the flour and 1/4 cup (50 mL) of sugar. Blend in the margarine. In a separate bowl beat the egg and almond extract, and mix in to make a smooth dough. Divide the dough in half and spread evenly over the bottom of 2 ungreased round layer pans.

Spread the Almond Filling over the top of each layer. Bake at 375°F (190°C) for 25 minutes. Cool for 5 minutes, then carefully remove layers from pan.

In a large bowl whip the cream, gradually adding 2 tablespoons (30 mL) of sugar and 1 teaspoon (5 mL) of almond extract.

Reserve 2 cups (500 mL) of whole strawberries. Slice the remaining berries. To assemble, place one crust layer on a serving platter. Cover with half the whipped cream and the sliced strawberries. Place the second layer over the top. Cover the top with the remaining whipped cream. Arrange the whole strawberries point up on top of the whipped cream.

Serves 8 to 10.

Strawberry Chocolate Torte

VERY QUICK AND ELEGANT!

1 pound cake, purchased
2 cups (500 mL) chocolate frosting
Fresh strawberries, sliced
½ cup (125 mL) whipping cream, whipped
Toasted sliced almonds

Slice the cake into 4 layers. Spread the frosting and a single layer of sliced strawberries between each layer of cake. Frost the sides and top of the cake. Spread whipped cream on top. Garnish with halved strawberries and toasted almonds. Chill until ready to serve.

Serves 6.

Strawberry Valentine Cake

Sponge Cake:
2 cups (500 mL) all-purpose flour
3 teaspoons (15 mL) baking powder
6 eggs, separated
2 cups (500 mL) sugar
1 teaspoon (5 mL) almond extract
1 cup (250 mL) warm water

Mallow Icing:
$\frac{1}{2}$ cup (125 mL) sugar
2 tablespoons (30 mL) water
2 egg whites
1 7-ounce (200g) jar marshmallow creme
$\frac{1}{2}$ teaspoon (2 mL) almond extract

1 cup (250 mL) sliced strawberries
10 to 12 whole strawberries for garnish

Grease and flour a heart-shaped pan In a medium bowl sift the flour and baking powder. In a separate bowl beat the egg whites until soft peaks form. Slowly beat in $\frac{1}{2}$ cup (125 mL) of sugar until stiff. Set aside.

In a large bowl beat the egg yolks until blended, then add the remaining $1\frac{1}{2}$ cups (375 mL) of sugar and the almond extract. Beat 4 minutes. Add the dry ingredients alternately with warm water. Fold in the beaten egg whites. Bake at 350°F (180°C) for 50 to 60 minutes or until done. Remove the cake from the pan before it is completely cooled. When cool, slice the cake in half horizontally.

In the top of a double boiler combine $\frac{1}{2}$ cup (125 mL) of sugar, water, and egg whites. Beat with electric mixer over boiling water until soft peaks form. Add the marshmallow creme and beat until stiff and glossy. Remove from the heat and beat in the almond extract. Spread a thin layer of mallow icing on the bottom half of the cake. Add a layer of sliced strawberries. Place the top half over the strawberries and frost with the remaining icing. Decorate with whole strawberries around the edge of the cake.

Serves 8.

Old World Strawberry Chocolate Cake

1½ cups (375 mL) sifted cake flour
1¼ cups (300 mL) sugar
⅓ cup (75 mL) cocoa, sifted
1 tablespoon (15 mL) instant coffee powder
1¼ teaspoons (6 mL) baking soda
½ teaspoon (2 mL) salt
⅔ cup (150 mL) margarine
1 cup (250 mL) buttermilk
1 teaspoon (5 mL) vanilla extract
2 eggs

2 tablespoons (30 mL) sugar
1 quart (1L) fresh strawberries, halved

Chocolate Glaze:
3 squares semisweet chocolate
3 tablespoons (45 mL) water
1 teaspoon (5 mL) instant coffee powder
1 tablespoons (15 mL) butter
1 cup (250 mL) confectioners sugar

2 cups sweetened whipped cream or whipped topping
Whole strawberries for garnish

Lightly oil two 8-inch (20 cm) round layer cake pans. In the bowl of an electric mixer combine the flour, 1¼ cups (300 mL) of sugar, cocoa, instant coffee powder, baking soda, and salt. Add the margarine, two-thirds of the buttermilk, and the vanilla. Beat on medium speed for 2 minutes. Add the remaining buttermilk and eggs. Beat 2 minutes more. Pour the batter into the prepared pans. Bake at 325°F (160°C) for 20 to 25 minutes.

Cool the layers on a wire rack.

In a large bowl sprinkle 2 tablespoons (30 mL) of sugar over the strawberries.

In a heavy saucepan melt the chocolate with the water, coffee powder, and butter over low heat. Remove the pan from the heat, and stir in the sugar. Add a little hot water if necessary to make a medium thin glaze.

Place one cake layer on a serving platter, and cover with half the strawberries. Drizzle with half the glaze and top with half the whipped cream. Repeat with the second layer. Garnish with whole strawberries.

Serves 8 to 10.

Strawberry Angel Cake

A COLORFUL, LIGHT DESSERT, AND SO SIMPLE TO PREPARE!

1 package angel food cake mix
1 3-ounce (85 g) package strawberry gelatin
1 cup (250 mL) boiling water
2 cups (500 mL) sliced strawberries
1 cup (250 mL) whipping cream, whipped

Prepare and bake the angel food cake according to the package directions. Cool. Tear the angel cake into bite-size pieces and set aside.

In a large bowl dissolve the gelatin in boiling water. Refrigerate until the gelatin is thickened but not set. Add the sliced strawberries. Fold in the stiffly beaten whipped cream. Place half the cake pieces in an angel food pan and pour half the berry mixture over the top. Repeat with the remaining cake pieces and berry mixture. Cover with plastic wrap and chill for several hours or overnight.

To serve, unmold and garnish with whipped cream and additional strawberries.
Serves 10 to 12.

STRAWBERRY SOUFFLÉ, PAGE 77

STRAWBERRY CHEESECAKE CUPS, PAGE 91

STRAWBERRY TIRAMISU, PAGE 104

STRAWBERRY ANGEL ROLL, 105

STRAWBERRY ANGEL CAKE, 112

PARISIENNE STRAWBERRY TART, PAGE 130

STRAWBERRY CHIFFON PIE, PAGE 140

STRAWBERRY ICE CREAM, PAGE 149

STRAWBERRY AND CHOCOLATE ICE CREAM TORTE, PAGE 151

STRAWBERRY BAKED ALASKA, PAGE 152

STRAWBERRY TEA, PAGE 161; STRAWBERRY MARGARITA, PAGE 162;
STRAWBERRY LOVING CUP, PAGE 160

PERFECT STRAWBERRY JAM, PAGE 168; STRAWBERRY JELLY, PAGE 169;
STRAWBERRY SYRUP, PAGE 170

PERFECT STRAWBERRY JAM, PAGE 168

STRAWBERRY IN THE GARDEN

STRAWBERRY FLOWER

BOWL OF STRAWBERRIES

Pies

and

Tarts

Parisienne Strawberry Tart

Tart Shell:

1½ cups (375 mL) all-purpose flour
¼ cup (50 mL) sugar
½ cup (125 mL) margarine
1 egg, beaten
1 tablespoon (15 mL) lemon juice

Filling:

4 cups (1 L) ripe strawberries, hulled
½ cup (125 mL) sugar
2 8-ounce (250 g) packages cream cheese, softened
½ cup (125 mL) sugar
2 teaspoons (10 mL) grated lemon rind
2 to 3 tablespoons (30-45 mL) light cream

Glaze:

1 tablespoon (15 mL) Amaretto liqueur
2 tablespoons (30 mL) cornstarch

In a large bowl combine the flour and ¼ cup (50 mL) of sugar. Cut in the butter until crumbly. Stir in the beaten egg and lemon juice to make a smooth dough. Chill.

On a floured board roll the dough to a ¼-inch (0.50 cm) thickness. Press into the bottom and half way up the sides of a springform pan. Bake at 400°F (200°C) for about 10 minutes or until golden. Cool.

Cut the large strawberries in half. Cut the small strawberries into slices. Place in separate bowls and sprinkle each with ¼ cup (50 mL) of sugar. Let stand 1 hour, to allow juice to form. Drain off the juice and set aside.

In a medium bowl beat the cream cheese until soft. Add ½ cup (125 mL) of sugar, the lemon rind, and enough cream to make a smooth mixture. Spread on the bottom of the cooled shell. Arrange the sliced berries over the top, then the halved berries, rounded side up.

Add enough water to the berry juice to make 1 cup (250 mL) of liquid. In a saucepan mix the Amaretto liqueur and cornstarch, then add the berry liquid. Cook and stir over medium heat until the mixture is thick and clear. Add 1 to 2 drops of red food coloring. Cool slightly and pour over the berries. Chill. Remove the sides of the pan to serve.

Serves 10 to 12.

Fresh Strawberry Pie

QUICK AND EASY!

2 cups (500 mL) fresh strawberries
1 tablespoon (15 mL) Cointreau
2 tablespoons (30 mL) sugar
1 tablespoon (15 mL) fresh lemon juice
Whipped cream
1 prepared graham cracker pie crust

Wash, hull, and slice the strawberries. Add the Cointreau, sugar, and lemon juice to the strawberries. Let stand for 15 minutes. Spread a thin layer of whipped cream on the bottom of the pie crust. Add the strawberries and juice. Top with the remaining whipped cream. Serve immediately.

Serves 6.

Strawberry Mallow Pie

Base:
1 cup (250 mL) chocolate wafer crumbs
¼ cup (50 mL) margarine, melted
Chocolate wafers, cut in half

Filling:
1 3-ounce (85 g) package strawberry gelatin
1 cup (250 mL) boiling water
1 cup (250 mL) whipping cream, whipped
2 cups (500 mL) miniature marshmallows
2 cups (500 mL) strawberries, slightly crushed

In a medium bowl combine the chocolate wafer crumbs and margarine. Press onto the bottom of a 9-inch springform pan or pie plate. Line the sides of the pan with wafer halves.

In a large bowl dissolve the strawberry gelatin in the boiling water. Chill until thick but not set. Fold in the stiffly whipped cream and marshmallows. Pour into the crust and chill until firm. Garnish with additional whipped cream and strawberries.

Serves 8 to 10.

Strawberry Cream Pie

Filling:
1 8-ounce (250 g) package cream cheese, softened
1 cup (250 mL) whipping cream
¼ cup (50 mL) sugar
1 teaspoon (5 mL) almond extract
1 ready-to-use graham cracker pie crust

Glaze:
3 cups (750 mL) strawberries, washed and hulled
1 tablespoon (15 mL) cornstarch
½ cup (125 mL) sugar
¼ cup (50 mL) water

In the bowl of an electric mixer beat the cream cheese, whipping cream, sugar, and almond extract on medium speed until fluffy and stiff peaks form. Spread over the graham crust, cover, and chill about 3 hours to set.

Crush enough of the strawberries to make ½ cup (125 mL).

In a saucepan combine the cornstarch, sugar, and water, and mix until blended. Add the crushed strawberries and cook over medium heat until the mixture thickens and comes to a boil. Boil for 1 minute. Cool. When the cream cheese filling is set, arrange the remaining whole strawberries on top. Pour the glaze over the strawberries and chill for about 1 hour until the glaze is set.

Serves 6.

Strawberry Almond Tart

3 egg whites
¼ teaspoon (1 mL) cream of tartar
¾ cup (175 mL) sugar
1 teaspoon (5 mL) almond extract
¾ cup (175 mL) slivered almonds, toasted

6 cups (1½ L) fresh strawberries

Glaze:
1 cup (250 mL) crushed strawberries
1 cup (250 mL) water
3 tablespoons (45 mL) cornstarch
¾ cup (175 mL) sugar

Topping:
½ cup (125 mL) light sour cream
1 to 2 tablespoons (15 to 30 mL) milk
Brown sugar

In a large bowl beat the egg whites and cream of tartar until soft peaks form. Add ¾ cup (175 mL) of sugar gradually, beating until stiff and glossy. Add the almond extract. Fold in the toasted almonds. Line a rectangular pan with baking paper and butter the sides and bottom well. Spread the meringue in the pan to form a shell. Bake at 300°F (150°C) for 50 to 60 minutes. Turn off the heat and leave in the oven for 1 hour. Cool.

Slice 5 cups (1¼ L) strawberries in half lengthwise and layer over the meringue. In a medium bowl crush the remaining berries.

To make the glaze, in a saucepan combine the crushed strawberries and with water. Bring to a boil and simmer for 2 minutes. In a small bowl combine the cornstarch and ¾ cup (175 mL) of sugar; and stir in a little of the berry mixture.Return the mixture to the saucepan. Cook over medium heat until thick and clear. Spread the glaze over the strawberries. Chill for 2 hours.

In a medium bowl combine the sour cream and milk, and drizzle over the strawberries. When ready to serve, sprinkle with brown sugar.

Serves 8.

Strawberry Chocolate Tart

Sweet Almond Dessert Pastry:
1¼ cups (300 mL) pastry flour
3 tablespoons (45 mL) sugar
½ cup (125 mL) cold butter
½ cup (125 mL) slivered almonds
1 egg, beaten
1 to 2 tablespoons (15 to 30 mL) cold water

Filling:
6 squares semisweet chocolate
2 tablespoons (30 mL) margarine
¼ cup (50 mL) sugar
3 tablespoons (45 mL) orange-flavored liqueur
1 8-ounce (250 g) package cream cheese, softened

4 cups (1L) fresh whole strawberries
2 cups (500 mL) whipping cream
2 tablespoons (30 mL) sugar

In a large bowl combine the flour and 3 tablespoons (45 mL) of sugar. Cut in the cold butter until crumbly. Mix in the slivered almonds. In a separate bowl combine the beaten egg and 1 tablespoon (15 mL) of water, and stir into the dry mixture with a fork to form a dough. If the dough seems too dry, add a little more water. Roll the dough out in a circle large enough to cover the bottom and half way up the sides of a 9-inch (22 cm) tart pan or springform pan. Prick the bottom of the pastry with a fork. Bake at 400°F (200°C) for 12 to 15 minutes or until golden brown. Cool.

In a medium saucepan melt the chocolate and margarine over very low heat. Add ¼ cup (50 mL) of sugar, the orange liqueur, and cream cheese. Stir until smooth and well blended. Remove from the heat. Cool at room temperature for 5 minutes.

Pour the chocolate mixture into the cooled pastry. Arrange the whole strawberries point up over the chocolate filling in the pastry shell. Chill completely.

When ready to serve, beat the cream and sweeten with 2 tablespoons (30 mL) of sugar. Top each serving with whipped cream.

Serves 8.

Strawberry Cloud Pie

AN ELEGANT DESSERT, AND SIMPLE TO MAKE!

Crust:
1½ cups (375 mL) graham cracker crumbs
¼ cup (50 mL) sugar
⅓ cup (75 mL) soft margarine

2 to 3 cups (500-750 mL) fresh strawberries

Topping:
1¼ cups (300 mL) sugar
½ cup (125 mL) water
½ teaspoon (2 mL) cream of tartar
3 egg whites
1 teaspoon (5 mL) almond extract

In a medium bowl mix together the graham cracker crumbs, sugar, and margarine, and press into a 9-inch (22 cm) pie plate. Bake at 375°F (190°C) for 8 minutes. Cool.

Wash, hull, and dry enough fresh strawberries to mound into the pie shell. You may leave the berries whole or slice them if preferred.

In a saucepan combine the sugar, cream of tartar, and water, and bring to a boil over medium heat. Cook uncovered until a small amount of the mixture forms a soft ball when placed in cold water. Beat the egg whites until very stiff. Add the almond extract and beat well. Pour the hot sugar mixture slowly into the beaten egg whites, beating constantly until thick and glossy. Spoon over the strawberries in the pie shell. Chill and serve.

Serves 6 to 8.

Strawberry Marzipan Tart

Tart Shell:
1¼ cups (300 mL) all-purpose flour
¼ cup (50 mL) sugar
⅓ cup (75 mL) soft margarine
1 egg yolk, beaten

Marzipan Layer:
5 tablespoons (75 mL) margarine
¾ cup (175 mL) almond paste
1 egg, beaten

Strawberry Filling:
2 cups (500 mL) fresh strawberries

Glaze:
1 cup (250 mL) red currant jelly

In a medium bowl combine the flour and sugar. Cut in the margarine with a pastry cutter until crumbly. Add the beaten egg yolk to make a smooth dough. Press the dough into the bottom and halfway up the sides of a 9-inch (22 cm) springform pan.

In a small bowl cream the margarine. Add the almond paste and beaten egg alternately until the mixture is smooth. Spread over the tart shell in the springform pan. Bake at 400°F (200°C) about 20 minutes until golden.

Cool completely.

Slice the strawberries and spread over marzipan layer in pan.

In a saucepan melt the jelly over low heat, and spoon evenly over the berries. Chill.
Serves 8.

Strawberry Chiffon Pie

2 cups (500 mL) flaked coconut
¼ cup (50 mL) margarine

2½ cups (750 mL) fresh strawberries
¼ cup (50 mL) sugar
1 tablespoon (15 mL) lemon juice

1 envelope unflavored gelatin
¼ cup (50 mL) sugar
¾ cup (175 mL) water

2 egg whites
¼ cup (50 mL) sugar
⅓ cup (75 mL) whipping cream

In a medium bowl mix the coconut and margarine and press into the bottom and up the sides of a pie plate to form a shell. Bake at 325°F (160°C) for 15 minutes or until golden. Cool completely.

In a bowl crush enough strawberries to measure 1½ cups (375 mL). Reserve the remaining berries for garnish. Stir ¼ cup (50 mL) of sugar and the lemon juice into the crushed berries, and let stand for 30 minutes to allow juices to form.

In a small saucepan mix the gelatin and ¼ cup (50 mL) of sugar together. Stir in the water. Heat and stir until the gelatin and sugar are dissolved. Cool. Stir the cooled gelatin mixture into the strawberry mixture. Chill until slightly thickened, stirring occasionally.

Remove from the refrigerator.

In a large bowl beat the egg whites to soft peaks. Gradually add ¼ cup (50 mL) of sugar, beating until stiff. Fold the beaten egg whites into the strawberry gelatin mixture. In a separate bowl beat the cream to soft peaks, and fold into the gelatin mixture. Pour into the cooled coconut crust. Chill.

Serve garnished with whipped cream and whole strawberries.

Serves 6 to 8.

Strawberry Tarts

½ cup (125 mL) heavy cream
1 envelope unflavored gelatin
¾ cup (175 mL) milk
1 10-ounce (450 g) package frozen strawberries, partially thawed
2 tablespoons (30 mL) sugar
½ teaspoon (2 mL) almond extract
6 prepared graham cracker tart shells
Whipped cream or topping

Pour the cream into a blender and sprinkle the gelatin over the top. Allow to stand for 5 minutes.

In a small saucepan heat the milk to boiling, and then add to the blender. Blend for 3 minutes or until the gelatin is dissolved. Add the strawberries, sugar, and almond extract, and blend on high speed until well blended. Chill until partially set.

Pour into tart shells and chill until set completely. Top with sweetened whipped cream or topping if desired.

Serves 6.

Fresh Strawberry Tartlets

Pastry:
¼ cup (50 mL) butter or margarine
1¼ cups (300 mL) all-purpose flour
2 tablespoons (30 mL) confectioners' sugar
1 egg yolk
5 tablespoons (75 mL) ice water

Pastry Cream:
1 cup (250 mL) light cream or milk
2 egg yolks
¼ cup (50 mL) sugar
2 tablespoons (30 mL) all-purpose flour
1 tablespoon (15 mL) butter
½ teaspoon (2 mL) vanilla extract

2 cups (500 mL) fresh small strawberries, washed
Confectioners' sugar

In a medium bowl combine the butter, 1¼ cups (300 mL) of flour, and the confectioners' sugar, and mix with a pastry cutter or food processor until coarse crumbs form. With a fork combine the egg yolk and water and add to dry mixture, mixing until the dough forms a ball. Roll the dough between two sheets of waxed paper until ⅛-inch (0.25 cm) thick. Using a 4-inch (10 cm) round cookie cutter, make tartlets and place in tartlet molds. Prick the bottom of the pastries with a fork. Chill for 25 minutes.

Bake at 400°F (200°C) for about 12 minutes until golden.

Cool slightly and then remove the pastries from the molds. Let cool completely.

In a small saucepan heat the cream over moderate heat just until simmering. Remove from the heat. In a large bowl combine the egg yolks, sugar, and 2 tablespoons (30 mL) of flour, and whisk together. Slowly whisk the hot cream into the egg yolk mixture. Transfer the mixture back to the saucepan and bring to a boil, whisking continuously for 2 minutes. Remove from the heat, and stir in the butter and vanilla. Transfer the hot cream mixture to a clean bowl and place plastic wrap directly on top the mixture to prevent a film forming. Allow to cool completely.

To serve, place 2 tablespoons (30 mL) of pastry cream into each tartlet shell. Top with strawberries, pointed end up. Dust lightly with confectioners' sugar.

Serves 10 to 12.

Strawberry Lemon Cream Tarts

QUICK AND WONDERFUL!

4 frozen individual puff pastry shells
½ cup (125 mL) whipping cream
⅓ cup (75 mL) lemon curd (purchased)
Fresh strawberries, washed, hulled, and halved lengthwise

Bake the shells according to the package directions. Cool.

In the bowl of an electric mixer whip the cream until stiff peaks form. Place the lemon curd in a bowl and fold in the whipped cream. Chill.

Spoon the lemon cream mixture into the pastry shells. Place halved strawberries cut side down on top. Garnish with whipped topping.

Serves 4.

Frozen

Desserts

Frozen White Chocolate Mousse Cake with Strawberries

Crust:
2 cups (500 mL) finely crushed vanilla wafers
1/3 cup (75 mL) melted butter

Mousse:
1/2 cup (125 mL) whipping cream
10 ounces (450 g) imported white chocolate
2 tablespoons (30 mL) almond liqueur
1/2 cup (125 mL) sugar
1/4 cup (50 mL) water
4 egg whites
1 1/2 cups (375 mL) chilled whipping cream

2 cups (500 mL) fresh strawberries

In a medium bowl mix together the vanilla wafers and butter, and press over the bottom and up the sides of a 9-inch (23 cm) springform pan. Bake at 350°F (180°C) for 8 minutes. Cool.

To prepare the mousse, in a saucepan heat the whipping cream over medium heat. Chop the chocolate into pieces and add to the hot cream. Reduce the heat to low and stir until smooth. Transfer the mixture to a large bowl and stir in the almond liqueur. Cool completely.

In a saucepan combine the sugar and water, and boil over high heat for 5 minutes.

While the syrup is boiling, in a bowl beat the egg whites until stiff peaks form. Slowly beat the boiling syrup into the egg whites. Beat continuously until stiff peaks form and the mixture is cool. Fold half of the meringue into the cooled white chocolate mixture, then fold in the remaining half.

In a separate bowl beat the chilled whipping cream until soft peaks form. Fold into the meringue and chocolate mixture. Pour the mousse mixture over the crust in the springform pan. Cover with plastic wrap and freeze until firm.

To serve, wash and hull the strawberries. Slice the berries in halves. Remove the mousse from the freezer and remove the sides of the pan. Arrange the strawberries over the top and serve immediately.

Serves 12.

Strawberry and White Chocolate Ice Cream with Strawberry Sauce

2 cups (500 mL) strawberry ice cream, softened slightly
3 ounces (85 g) white chocolate, grated

Strawberry Sauce:
2 cups (500 mL) strawberries, sliced
2 tablespoons (30 mL) sugar
1/2 teaspoon (2 mL) almond extract

Spoon the softened ice cream into a bowl. Mix in the grated white chocolate. Cover and freeze.

In a blender combine the sliced strawberries, sugar, and almond extract, and purée until smooth. Chill.

To serve, scoop the ice cream into dessert glasses and top with strawberry sauce.

Serves 6 to 8.

Strawberry Ice Cream

3 egg yolks
4 cups (1 L) whipping cream
1 cup (250 mL) sugar
1 or 2 tablespoons (15-30 mL) strawberry liqueur (optional)
1 cup (250 mL) fresh strawberries, sliced

In a large bowl beat the egg yolks. Add the whipping cream, sugar, and liqueur. Mix until the sugar is dissolved. Pour into a glass bowl or pan and place uncovered in the freezer. Stir occasionally to keep ice crystals from forming. Freeze until slushy.

Stir in the strawberries. Place in a covered container and freeze until firm. Store in the freezer.

If you have an ice cream machine, simply add all of the ingredients and follow the instructions for the machine.

Makes about 6 cups (1½ L).

Strawberry Sherbet

4 cups (1 L) fresh strawberries
2 cups (500 mL) buttermilk
¼ cup (50 mL) liquid honey

In a food processor or blender chop the strawberries. Stir in the buttermilk and honey. Place in ice cube trays (dividers removed), cover with foil, and freeze until slushy. Remove from the freezer; return to the blender or processor, and process until smooth and creamy. Do not allow the mixture to melt. Return to the freezer trays and freeze. Stir occasionally. Freeze 4 hours or until completely frozen.

If you have an ice cream machine, simply add all of the ingredients and follow the instructions for the machine.

To serve, allow to soften slightly in the refrigerator for 15 to 20 minutes.

Serves 8.

Strawberry and Chocolate Ice Cream Torte

1 18¼-ounce (525 g) package white cake mix
Chunky chocolate ice cream
Strawberry ice cream
2 cups (500 mL) whipping cream
¼ cup (50 mL) sugar

In a large bowl mix the cake batter according to the package directions. Pour the batter into a greased 9 x 5-inch (23 x 12 cm) loaf pan. Bake at 350°F (180°C) for 50 to 60 minutes or until the cake tests done.

Cool for 15 minutes and remove from the pan. When completely cool, slice the cake into 3 layers. Spread a layer of softened chocolate ice cream over the bottom cake layer. Place the second cake layer over the top. Spread a layer of softened strawberry ice cream on the second layer. Top with the third cake layer. Wrap and freeze until completely firm.

To serve, whip the cream until stiff, adding sugar to sweeten. Cover the tops and sides of the frozen torte. Garnish with strawberries. Serve at once.
Serves 8.

Strawberry Baked Alaska

THIS IS VERY ELEGANT AND DELICIOUS, YET SIMPLE TO MAKE.

Crust:
1½ cups (375 mL) graham cracker crumbs
¼ cup (50 mL) sugar
¼ cup (50 mL) soft margarine

Filling:
2 cups (500 mL) strawberries, sliced
Vanilla or strawberry ice cream, slightly softened

Meringue Topping:
3 egg whites
¼ cup (50 mL) sugar
1 teaspoon (5 mL) almond extract

In a medium bowl mix the graham cracker crumbs, ¼ cup (50 mL) of sugar, and margarine together. Press the crumb mixture into a 9-inch (22 cm) pie plate to form a crust. Bake at 375°F (190°C) for 8 minutes.

Cool completely, then chill in the freezer.

Place a layer of strawberries on the bottom of the crust. Spoon softened ice cream evenly over top and add another layer of strawberries. Cover with plastic wrap and return to the freezer while preparing the meringue.

Prepare the meringue just before serving. Preheat the oven (on broiler setting) to 500° F (260°C). In the bowl of an electric mixer beat the egg whites until soft peaks form. Slowly add ¼ cup (50 mL) of sugar, beating until very stiff and glossy. Beat in the almond extract.

Spread the meringue over the strawberries and ice cream. Be sure to bring the meringue to the edge of the crust to seal. Place under the broiler, just until golden. Watch carefully! Serve at once.

Serves 6 to 8.

Strawberry Ice Cream Roll

⅔ cup (150 mL) cake flour
1 teaspoon (5 mL) baking powder
4 eggs
¾ cup (175 mL) sugar
2 tablespoons (30 mL) margarine, melted
Confectioners' sugar
2 cups (500 mL) strawberry or vanilla ice cream, slightly softened
2 cups (500 mL) ripe strawberries, coarsely chopped

Line a jelly-roll pan with waxed paper, grease well, and set aside. In a medium bowl combine the flour and baking powder, and set aside.

In a large bowl beat the eggs and sugar with an electric mixer at high speed for about 10 minutes until thick and foamy. Fold in the flour mixture and melted margarine. Pour into the pan, spreading evenly. Bake at 400°F (200°C) for 8 to 10 minutes.

Invert onto a clean towel sprinkled with confectioners' sugar. Remove the paper and roll up lengthwise with a towel. Cool.

When ready to serve, carefully unroll, removing the towel. Trim the crisp edges. Quickly spread with softened ice cream and sprinkle with chopped strawberries. Starting at a long side, roll up. Cut into slices and serve at once. This dessert freezes well and may be prepared ahead. Wrap well before freezing.

Serves 10.

Frozen Desserts

Frozen Strawberry Parfait

¾ cup (175 mL) sugar
½ cup (125 mL) water
4 egg yolks
4 cups (1 L) fresh strawberries, washed and hulled
2 cups (500 mL) whipping cream
¼ cup (50 mL) confectioners' sugar

Lightly oil an 8-cup (2 L) decorative mold, and chill in the freezer while preparing the dessert.

In a saucepan mix the sugar and water, and bring to a boil. Stir until the sugar is dissolved. Continue to boil without stirring until the syrup spins a "thread" when dropped from a spoon, about 5 minutes. In a small bowl beat the egg yolks until thick and lemon colored. Slowly add the hot syrup mixture, beating constantly for 5 minutes. Set the bowl in ice water for 10 minutes to cool, stirring constantly.

In a blender purée the strawberries. In a chilled bowl beat the whipping cream and confectioners' sugar until stiff. Mix the strawberry purée and egg yolk mixture together, and fold into the whipped cream. Pour into the mold, and cover with foil. Freeze overnight.

To serve, unmold onto a serving platter. Shake gently to release. A hot damp dish towel placed over the mold will help to loosen. Garnish with whole strawberries placed around the dessert.

Serves 12.

Strawberry Freeze

Base:
1 cup (250 mL) all-purpose flour
¼ cup (50 mL) firmly packed brown sugar
½ cup (125 mL) chopped pecans
½ cup (125 mL) margarine, melted

2 egg whites
½ cup (125 mL) sugar
1 cup (250 mL) whipping cream, whipped
1 tablespoon (15 mL) lemon juice
2 cups (500 mL) sliced fresh strawberries (or frozen strawberries,
thawed and drained)

In a medium bowl combine the flour, brown sugar, and pecans. Add the melted butter and mix until crumbly in texture. Spread into an 8-inch (20 cm) square pan. Bake at 350°F (180°C) for 25 minutes to toast, stirring occasionally. Let cool. Remove ⅓ cup (75 mL) for topping.

In a large bowl beat the egg whites until soft peaks form. Gradually add the sugar and beat until very stiff. Fold in the whipped cream. In a medium bowl mix the lemon juice and strawberries, and fold in quickly. Spread over the toasted crumb base and top with the remaining crumbs. Freeze until firm. Remove from the freezer 20 minutes before serving. Cut into squares.

Serves 9.

Frozen Strawberry Yogurt Pie

Crust:
1/3 cup (75 mL) wheat germ
1 cup (250 mL) quick-cooking oats
1/4 cup (50 mL) firmly packed brown sugar
1 teaspoon (5 mL) ground cinnamon
3 tablespoons (45 mL) margarine, melted
1/4 cup (50 mL) chocolate chips

Filling:
4 cups (1 L) frozen strawberry yogurt, softened
2 cups (500 mL) sliced strawberries

In a medium bowl combine the wheat germ, oats, brown sugar, cinnamon, and margarine, and press into the bottom and sides of a 9-inch (22 cm) pie pan. Bake at 375°F (190°C) for 8 minutes.

Remove from the oven and sprinkle the bottom of the crust with chocolate chips. Cool completely.

Spoon the softened yogurt into the cooled crust and spread evenly. Freeze until firm. To serve, top the pie with sliced strawberries.

Serves 6 to 8

Beverages
and
Ices

Wild Strawberries in Champagne

IF YOU CAN FIND WILD STRAWBERRIES FOR THIS,
YOU WILL HAVE THE MOST BEAUTIFUL PUNCH IN THE WORLD!!

2 cups (500 mL) ripe strawberries
½ cup (125 mL) fine sugar
2 tablespoons (30 mL) fresh lemon juice
2 tablespoons (30 mL) Kirsch
1 bottle chilled Champagne

Rinse the strawberries in cold water. Drain on a paper towel. Place in a crystal bowl, and sprinkle with sugar, lemon juice, and Kirsch. Chill for 3 to 4 hours. When ready to serve, pour the chilled champagne over the berries. Stir. Serve in champagne glasses with a strawberry.

Serves 4.

Strawberry Frappé

DELICIOUSLY COOL!

2 cups (500 mL) fresh strawberries
1 cup (250 mL) club soda, chilled
Sugar to taste
Crushed ice
1 fresh lime

Wash and hull the strawberries. In a blender purée the strawberries until smooth. Add the club soda and sugar. Stir in the crushed ice.

To serve, spoon the strawberry slush into attractive crystal wine or champagne glasses. Garnish with a lime wedge or whole strawberry.

Serves 4.

Strawberry Loving Cup

1½ cups (375 mL) milk
6 fresh strawberries
3 ounces (75 mL) Amaretto liqueur
Sugar to taste
Whipped cream
Toasted, crushed almonds

In a blender combine the milk and strawberries, and blend until smooth. Transfer to a saucepan and add the Amaretto and sugar. Heat only until warm, do not boil.

Garnish with sweetened whipped cream and toasted almonds.
Serves 2 generously.

Strawberry Tea

1 cup (250 mL) blush or rosé wine
½ cup (125 mL) strong tea
½ cup (125 mL) orange juice
½ cup (125 mL) fresh strawberries
Sugar or sweetener to taste

In a blender combine all of the ingredients. Blend together on high speed just until well mixed.

Serve with ice cubes in tall glasses.
Serves 2.

Strawberry Marguerita

2 cups (250 mL) fresh ripe strawberries
8 ice cubes, crushed
Dash fresh lime juice
Tequila

In a blender combine all of the ingredients. Blend on low speed. The amount of tequila used is a matter of preference, but 2 ounces works well in this recipe.

Serve in chilled, salt-rimmed cocktail glasses. Garnish with whole strawberries.
Serves 2.

Strawberry Sodas

FOR EACH SERVING:

¼ cup (50 mL) strawberry syrup (see recipe, page 170)
¼ cup (50 mL) milk
2 to 3 scoops strawberry ice cream
Club soda, chilled

In a tall ice cream soda glass combine the strawberry syrup, milk, and ice cream. Fill with club soda. Garnish with a whole strawberry.
Serves 1.

Strawberry Yogurt Shake

2 cups (500 mL) plain yogurt
1 cup (250 mL) ripe strawberries
Sugar or sweetener to taste

In a blender combine the yogurt and strawberries. Blend on medium speed until smooth and frothy. May be sweetened to taste if preferred.

Pour into tall glasses.
Serves 2.

Strawberry Nog

NUTRITIOUS AND REFRESHING.

1 cup (250 mL) plain yogurt
1 cup (250 mL) whole, 2%, or skim milk
1 cup (250 mL) sliced fresh strawberries
4 teaspoons (20 mL) sugar
1 egg
½ teaspoon (2 mL) vanilla extract
2 to 3 ice cubes

In a blender combine the yogurt, milk, strawberries, sugar, egg, and vanilla extract. Blend until frothy. Add the ice cubes one at a time, blending each time. Serve in tall glasses.
Serves 2 generously.

Beverages and Ices

Strawberry Buttermilk Cooler

2 cups (500 mL) cold buttermilk
1 cup (250 mL) sliced, fresh strawberries
Sugar to taste
2 ice cubes

In a blender combine all of the ingredients. Cover and blend on high speed until smooth. Serve in glasses.
Serves 2.

Wild Strawberry Ice

2 cups (500 mL) water
1 cup (250 mL) sugar
1 quart wild strawberries, washed and hulled

In a saucepan combine the water and sugar and bring to a boil over high heat. Reduce the heat and allow to simmer for 10 minutes to make a syrup.

Remove the pan from the heat. Add the wild strawberries and allow to cool. Strain the strawberries and syrup mixture through a fine sieve into a bowl. Freeze just until slushy, stirring occasionally as ice crystals form.

Serve in chilled champagne glasses.
Serves 4 to 6.

May Wine with Wild Strawberries

PERFECT FOR A WEDDING CELEBRATION!

3 pints (1½ L) ripe wild strawberries
½ cup (125 mL) sugar
2 bottles red or blush wine
1 bottle Champagne, chilled
Fresh rose petals, apple blossoms, or violets, washed
Reserved whole wild strawberries

In a large bowl crush the strawberries and stir in the sugar. Add the wine and stir to dissolve the sugar. Cover and chill for 24 hours.

To serve, transfer the strawberry-wine mixture into a punch bowl. Add the champagne and stir. Garnish with rose petals, apple blossoms, or violets and reserved wild strawberries.

Makes about 3 quarts (3 L).

Strawberry Sangria

1 cup (250 mL) strawberries, sliced
¼ cup (50 mL) sugar
Juice of 1 orange and 1 lime
1 bottle rosé wine, chilled
Thin orange and lime slices
Whole strawberries, halved

In a blender combine the sliced strawberries, sugar, orange juice, and lime juice. Purée until smooth. Transfer to a glass pitcher and add the rosé wine. Stir well. Add thin slices of orange and lime. Stir in the halved strawberries. Chill for several hours to allow the flavors to blend.

Serve over crushed ice or Strawberry Ice Cubes (recipe follows). Recipe may be easily doubled.

Makes 6 cups (1½ L).

Strawberry Ice Cubes

AN ELEGANT TOUCH FOR ALL COLD BEVERAGES!

Wash small ripe strawberries, leaving the hull on for color. Place the berries in ice cube trays, and add water to cover the berries. Freeze.

Add strawberry ice cubes to all cold drinks. It's a colorful and refreshing touch.

Note: You may also want to try this with other berries or cut fruit pieces. Small edible flowers from your garden and flower petals from larger flowers add an especially nice touch. Just wash thoroughly before using.

Jams

and

Jellies

Perfect Strawberry Jam

GOOD OLD FASHIONED JAM MADE WITH FRESH STRAWBERRIES!

4 cups (1 L) fresh ripe strawberries
7 cups (1.75 L) sugar
3 ounces (85 mL) liquid pectin

Wash and hull the strawberries. Slice the large strawberries lengthwise. Halve the smaller berries. Add 1 cup (250 mL) of sugar to the berries and stir well. Allow to stand for 15 minutes to allow juice to form.

Place the strawberries in a large nonaluminium pot and stir in the remaining sugar. Bring to a hard, rolling boil over medium-high heat. Boil for 1 minute, stirring constantly. Remove from the heat and add the liquid pectin. Stir and skim the foam from the surface for 5 minutes. Ladle the hot jam into hot sterilized jars, leaving ½ inch (1 cm) at the top of the jar. Seal with melted paraffin wax. Screw metal lids tightly in place.

Store in a cool, dark place. Use within a few months.

Makes seven 8-ounce (250 mL) jars.

Strawberry Jelly

A DELICATE, BLUSH-COLORED JELLY.

1½ envelopes unflavored gelatin
½ cup (125 mL) sugar
½ cup (125 mL) fresh orange juice
2 cups (500 mL) fresh strawberries, puréed

In a small saucepan combine the gelatin, sugar, and orange juice over medium heat. Stir until the gelatin and sugar are dissolved. Transfer to a bowl and add the puréed strawberries. Stir to blend well. Ladle the hot jelly into hot sterilized jelly jars. Seal with melted paraffin wax. Screw metal lids in place.

Store in the refrigerator. Use within 2 weeks.

Makes three 8-ounce (250 mL) jars.

Strawberry Syrup

A WONDERFUL TOPPING FOR ICE CREAM AND DESSERTS,
OR USE SYRUP TO FLAVOR MILK SHAKES AND SODAS.

3 cups (750 mL) fresh strawberries
⅔ cup (150 mL) sugar
⅔ cup (150 mL) water

Wash and hull the strawberries. Slice the strawberries and purée in a blender. Transfer to a stainless steel saucepan and stir in the sugar and water. Bring to a boil over medium heat, and cook for 1 minute. Remove from the heat and force the mixture through a fine sieve set over a bowl. Cover and chill.

Keeps for one week in the refrigerator.

Makes approximately 3 cups (750 mL).

INDEX